EXPANDING THE SCOPE
OF SOCIAL SCIENCE
RESEARCH ON DISABILITY

EXPANDING THE SCOPE OF SOCIAL SCIENCE RESEARCH ON DISABILITY

Edited by   BARBARA M. ALTMAN
Agency for Health Care Policy and Research
Gallaudet University

SHARON N. BARNARTT
Agency for Health Care Policy and Research
Gallaudet University

Copyright © 2000 by JAI PRESS INC.
100 Prospect Street
Stamford, Connecticut 06901-1640

All rights reserved. No part of this publication may be reproduced, stored on a retrieval system, or transmitted in any form or by any means, electronic, mechanical, photocopying, filming, recording, or otherwise, without prior permission in writing from the publisher.

ISBN: 0-7623-0551-7

Manufactured in the United States of America

ACKNOWLEDGMENT

The picture on the cover is a photograph of porcelain pieces from the Lunar Eclipse Series created by Jacqueline Clipsham in 1982. Ms. Clipsham is a ceramic artist with a disability who has a studio in Califon, New Jersey. Photo by Hiro Ihara.

CONTENTS

LIST OF CONTRIBUTORS   vii

INTRODUCING RESEARCH IN SOCIAL SCIENCE
AND DISABILITY: AN INVITATION TO SOCIAL
SCIENCE TO "GET IT"
Barbara M. Altman and Sharon N. Barnartt   1

IDIOT INTO APE
Chris Borthwick   31

HOW THE NEWS FRAMES DISABILITY: PRINT
MEDIA COVERAGE OF THE AMERICANS WITH
DISABILITIES ACT
Beth Haller   55

GENDER CONTRADICTIONS AND STATUS
DILEMMAS IN DISABILITY
Judith Lorber   85

SUCCESSFUL LABOR MARKET TRANSITIONS FOR
PERSONS WITH DISABILITIES: FACTORS
AFFECTING THE PROBABILITY OF ENTERING
AND MAINTAINING EMPLOYMENT
Edward Yelin and Laura Trupin   105

RESEARCH ON HEALTH CARE EXPERIENCES OF
PEOPLE WITH DISABILITIES: EXPLORING THE
SOCIAL ORGANIZATION OF SERVICE DELIVERY
Marie L. Campbell   131

DISABILITY-RELATED INTENTIONAL INJURY
HOSPITALIZATIONS: A MULTI-STATE ANALYSIS
Llewellyn J. Cornelius   155

DISENFRANCHISED: PEOPLE WITH DISABILITIES
IN AMERICAN ELECTORAL POLITICS
  Todd G. Shields, Kay Schriner, Ken Schriner, and Lisa Ochs   177

STADIUM SIGHTLINES AND WHEELCHAIR
PATRONS: CASE STUDIES IN IMPLEMENTATION
OF THE ADA
  Sanjoy Mazumdar and Gilbert Geis   205

LIST OF CONTRIBUTORS

Chris S. Borthwick   School of Health Sciences
LaTrobe University

Marie L. Campbell   Human and Social Development
University of Victoria

Llewellyn J. Cornelius   School of Social Work
University of Maryland at Baltimore

Gilbert Geis   Department of Criminology, Law and Society
University of California, Urvine

Beth Haller   Department of Mass Communication
Towson University

Judith Lorber   Graduate School and Brooklyn College
City University of New York

Sanjoy Mazumdar   Department of Urban and Regional Planning
University of California, Irvine

Lisa Ochs   Department of Rehabilitation Education and Research
University of Arkansas

Kay Schriner   Department of Political Science
University of Arkansas

Ken Schriner Department of Political Science
University of Arkansas

Todd G. Shields Department of Political Science
University of Arkansas

Laura Trupin Department of Medicine and
Institute for Health Policy Studies
University of California,
San Francisco

Edward Yelin Department of Medicine and
Institute for Health Policy Studies
University of California,
San Francisco

INTRODUCING RESEARCH IN SOCIAL SCIENCE AND DISABILITY
AN INVITATION TO SOCIAL SCIENCE TO "GET IT"

Barbara M. Altman and Sharon N. Barnartt

This new volume series is launched with the purpose of stimulating, providing a vehicle for and encouraging the social sciences to address the many social issues associated with disability. Many changes in technology and social relations have made human disability more inevitable and less ignorable. More than ever before, children, adults and elderly populations experience chronic health conditions, accidents, and are exposed to war and environmental toxins with resulting damage to mind, body and/or spirit. Medical structures have expanded and taken responsibility for those in the population who experience disease or injury that results in disability. However, since people with disabilities both need and want to function within the society as a whole, medical knowledge, technology and institutions are inadequate to address issues that take place in the social realm. This is the province of social science disciplines, which have been as blind to the social, historical political, psychological and economic issues associated with disability as medicine has been.

While there is a movement toward the development of a disability studies discipline, not unlike women's studies or African American studies, the humanities have been more successful in establishing research and curriculum in these areas than the social sciences (Monaghen 1998). There is currently no journal which is dedicated to approaching the study of disability using just the theories, methods and perspectives of the social science disciplines. Established journals in the social science disciplines seldom publish articles on disability-related topics. This appears to be because disability is not widely recognized as a phenomenon which is both socially defined and which has enormous consequences for individuals and societies.

Refereed journals which do publish articles on disability are usually oriented to medical, rehabilitation or special education perspectives. A few journals do exist which focus on one aspect of disability research or one discipline. These include the *Journal of Disability Policy Studies* which, although multidisciplinary, focuses on disability policy at either the macro or micro levels; *Disability and Society*, a sociological journal that is predominantly theoretical; and *Disability Studies Quarterly* which combines short, non-peer reviewed articles, book and film reviews and recent disability news and information.

In this paper we attempt to give a brief introduction to the potential the topic of disability holds for social science. This paper is divided into three sections. In the first, we discuss the evolution and challenge of defining disability and offer a short overview of social science research about disability over the past 30 to 40 years. In the second section, we identify new areas for research in disability which have had little or no serious attention from social science disciplines. We also attempt to delineate some of the perspectives by which social science disciplines can approach these and other disability questions. Finally we give an overview of the papers that are included in this first volume of *Research in Social Science and Disability*, highlighting the new and insightful contributions they make to the field.

OVERVIEW OF DISABILITY DEFINITION AND SOCIAL SCIENCE RESEARCH IN DISABILITY

Conceptualizations and Definitions

One of the areas of disagreement in the evolution of studies of disability is what, in fact, constitutes a disability. In this paper and this series, we will follow terminology originally suggested by Nagi (1965), which served as a basis for examination and elaborations that resulted in two reports issued by the Institute of Medicine (IOM) of the National Academy of Sciences, (Pope and Tarlov 1991; Brandt and Pope 1997). In the more recent of these, the IOM panel strongly endorsed a conceptualization of disability which incorporates environmental factors as a primary element in creating disability. It recommended shifting the focus of the definition of disability from the individual and the impairment to the interaction between impairment and environment. This concept challenges the medical model, which sees disability as primarily a medical condition in need of remediation. In this view, the cultural category of "disabled" is socially, rather than medically, constructed, in part by cultural definitions and in part by the demands and limitations of the social and physical environments.

Based upon this concept of disability, the IOM report distinguished between a pathology, an impairment, a functional limitation, a potentially disabling condition and a disability. An active pathology, which can be an infection, a metabolic imbalance or traumatic injury, occurs at the biological level, and may change the body in some way. Examples are influenza, a breast tumor or a broken bone. An impairment is residual damage or loss, usually resulting from a pathology, which occurs at the level of the organ system or body part. Examples could include the heart, damaged by a heart attack; a leg, damaged by an accident; or the auditory nerve, damaged by long exposure to rock music. A functional limitation is an inability, arising from an impairment, to perform basic human functions such as walking or hearing. A strict interpretation of the IOM model would not consider dressing or making a phone call as functional limitations, because these are tasks associated with social roles and entail the interaction of people with their environments. A chronic or permanent functional limitation is a potentially disabling condition. For example, a child who is audiologically classified as being deaf would be considered to have a potentially disabling condition. In the IOM model, a potentially disabling condition is not seen as the equivalent of a disability until interactions with the person's physical and social environments are taken into consideration. Disability is a limitation in ability to perform parts or the whole of a desired social role, such as working, being a parent or attending church, in the physical and social environments where these activities take place. While there are some types of impairments that might make it impossible for a person to perform any social role in any environment, there are not many. Whether or not an impairment becomes disabling is determined by a combination of the limitations caused by the impairment, the accessibility of the environment (both social and physical) and the requirements of the role.

Thus, for example, if all of the family members of the deaf child are also deaf, and all communicate using American Sign Language (ASL), the child is not disabled within the family environment. The child is only disabled when (or if) she or he begins to attend a school in which others do not use sign language. It is that specific environment which places limitations on the child's ability to perform adequately or optimally in the student role. In a society in which "everyone speaks sign language" (Groce 1985), people with functional limitations in their auditory systems—whom society would consider to be deaf—are not disabled. Rather, they have a potentially disabling condition which may become disabling in other environments.

Another example would be a person who has a spinal cord injury which requires the use of a wheelchair. That person would have a work disability if employed in the manual aspects of construction but not if employed as a professor (if all of the buildings were wheelchair accessible, all sidewalks had curb cuts and lowered desks and blackboards were available).

The IOM conceptualization of disability implies that a person cannot be categorized as "disabled" without reference to the environment in question. A person might not be disabled in the family environment but might be in the work environment. Another person with the same physical impairment might be disabled in the family environment but not in the work environment. Thus, knowledge of a medical diagnosis alone cannot be used, a priori, to determine whether or not the person can be categorized as "disabled." Nor can it indicate the level or extent of disability the person would experience. In this concept, disability becomes a socially determined condition. As such, this concept most nearly fits the social science perspective (if the social sciences can be said to have one perspective) and is most appropriately used in this context.[1]

Disability Research in the Social Sciences

Although Nagi and his colleagues made a significant contribution to the sociology of disability as early as 1965, as the result of a conference and publication of the proceedings, much of the work of social scientists in this field has found limited exposure, particularly within social science venues. Journals focused specifically on social science perspectives on disability did not appear until the late 1980s or early 1990s. *Disability, Handicap and Society* (now known as *Disability and Society*) had its first publication in 1986, followed by *Journal of Disability Policy Studies* in 1990. They represented sociological and political science perspectives, respectively, but have drawn from multiple social science disciplines. Prior to the inauguration of these journals, social science examinations of disability issues found outlets mostly in medical journals; in journals related to specific disabilities, such as the *American Annals of the Deaf*, the *Journal of Visual Impairment and Blindness*, or the *American Journal of Mental Retardation*; or in books or edited volumes (Crammette 1968; Davis 1963; Croog and Levine 1977; Finlayson and McEwan 1977; Albrecht 1976).

Limited by the medical model and its equating of disability with dependency, predisposing the need for welfare and other forms of social insurance, much of the research that was produced between the late 1960s and early 1980s was focused on individuals and their adjustment to the dependency status. Psychologists, social psychologists and sociologists were concerned about psychological adjustment to, and coping with, the impairment (Kelman et al. 1964; Cohn 1970; Ludwig and Collette 1970; Ben-Sira 1981, 1983; Safilios-Rothchild 1970; Shontz 1975; Wright 1960); morale and motivation in rehabilitation settings (Barry et al. 1968; Starkey 1968; Brown and Rawlinson 1976; Cassem and Hackett 1973; Litman 1966); and levels of social support in the family and community (New et al. 1968; Tolsdorf 1976; Smith 1979, 1981; Peterson 1979). Another area, examined extensively, which *appeared* to touch on environmental components of disability in this early period was the study of the attitudes of peers, employers and others toward persons with disabilities (Yuker et al. 1960, 1966; Siller and Chipman 1964; Shears and Jensema 1969; Richardson 1968, 1970,

1971; Schroedel 1978). However, the lack of a truly environmental perspective in these early studies was captured by a critique of the attitude literature by Altman (1981).

At the same time that sociologists and psychologists were primarily concerned with the individual with a disability, economists were examining national issues associated with disability benefits and employment. Monroe Berkowitz, an early contributor to this field, focused on the evaluation of the structures and functions of disability programs, both workman's compensation and Social Security (Berkowitz 1973; Berkowitz et al. 1979; Berkowitz et al. 1971). Other issues economists focused on were the economic costs of vocational rehabilitation or of particular types of disability such as mental retardation (Conley 1965, 1973), analyses of the Social Security system and other public programs directed at persons with disabilities (O'Neill 1976; Worrall 1978; Johnson 1979; Meer 1979; Peck 1983) and the effects of disability on the labor supply (Swisher 1973; Scheffler and Iden 1974; Yelin et al. 1980; Slade 1984). The population of focus for economists was somewhat narrower than for sociologists and psychologists. Identification of the disabled population was very much tied to work status and the receipt of benefits such as SSDI.

The passage of the Rehabilitation Act of 1973, which used a civil rights perspective in its view of disability, presaged a movement in the literature to a minority perspective toward disability in the 1980s (Christiansen and Barnartt 1987; Gliedman and Roth 1980; Stroman 1982; Hahn 1983, 1985). Both views helped raise questions about the medical model of disability, with its emphasis on the individual and the biophysiological components of disability by focusing attention on the physical and social environment (DeJong 1979; DeJong and Wenker 1983; Hahn 1982).

Paralleling this development in the United States of conceptualizing persons with disabilities as a minority, work was also taking place in England. An attempt to create a classification scheme to define and measure disability had been initiated in the 1970s, culminating in what has become known as the ICIDH, the International Classification of Impairment, Disability and Handicap (WHO 1980). Primarily a classification scheme based on the work of Wood (1980), the ICIDH was the first to try more explicitly to incorporate the environmental effects into a conceptual model of disability. Paralleling the developments of this classification scheme, groups of persons with disabilities were organizing into the Disability Alliance and the Union of the Physically Impaired Against Segregation (UPIAS) supporting work which emphasized theoretical perspectives, particularly Marxist perspectives (Oliver 1990) and later cultural perspectives (Shakespeare 1994). This work took the form of the development of a social model of disability that sought to sever the relationship to medicine by ignoring the individual functional limitations and focusing totally on what was interpreted as an oppressive environment and social structure (Abberley 1987; Oliver 1990).

There had been signs of a broader perspective sprinkled throughout some social science literature starting with Bogdan and Biklen's examination of handicapism (1977), followed by Albrecht and Levy's framing of disability as a social problem (1981), Altman's critique of attitude literature (1981) and Hahn's explanation of the paternalistic nature of public policy (1983). Other defining models began appearing by this point, moving away from what was considered a totally medical orientation to incorporate the environmental element more specifically than in the past (Altman 1984; Verbrugge 1990; Verbrugge and Jette 1994). However, the inclusion of the environment at the early point was seen either as an epidemiological factor contributing to the severity of disability (Wan 1974) or as the disadvantage which the disability created for the individual in the social context. (Wood 1980). A little later, Verbrugge (1990), Badley (1987) and Altman (1984) began to define disability as the outcome or consequence of impairment and functional limitations influenced by the medical and social context in which they took place.

However, it is in the last decade that we have seen a burgeoning of new research in certain areas including demography and epidemiology (LaPlante 1991a, 1991b, 1993; LaPlante and Carlson 1996; Ing and Tewey 1994; Kaye et al. 1996), conceptualizations, definitions and measures (Brown 1990; Pope and Tarlov 1991; Verbrugge and Jette 1993; Altman 1993), theory analysis (Shakespeare 1994; Neath 1997) and political (Campbell and Oliver 1996). Analytical research has also moved away from purely health and policy areas, expanding into topics rarely examined earlier such as voting behavior (Schriner et al. 1998; Shields et al. 1998), abuse and violence (Nosek et al. 1997) and social movements (Barnartt and Scotch forthcoming). This increase in social science attention to disability theory, methodology and issues research is long overdue but encouraging. We would like to facilitate this broader interest by challenging more than a small core of persons in social science to consider the interesting questions that are associated with ways that people with impairments live in society. In the section that follows, we outline just a few of the issues that would benefit from social science research.

CURRENT CHANGES CREATE NEW OPPORTUNITIES

Many aspects of disability are changing in ways that open up new research possibilities for social scientists. Although addressing some of these situations will require applied research, basic theoretical and empirical work in the social sciences will also be necessary. Some examples of the current changes taking place and researchable issues associated with them are as follows:

1. Increasing population size, accompanied by increasing longevity, has resulted from the increased ability of medicine to prolong life in persons with serious diseases and injuries. This has also led to increases in the actual number of persons with a wide range of potentially disabling conditions. Increased pollution, violence, military actions and continuing socioeconomic inequality also contribute to the potential levels of disability. It is likely that the widely held belief that disability is a "one-in-ten" phenomenon will have to be re-examined both nationally and internationally.

There is ambiguity in the epidemiological and demographic literature on actual population proportions, particularly in less-developed countries. It is not clear if the number of persons with disabilities is increasing more in the developed world than the less-developed world. Demographic and other types of research are needed which would focus on this and related questions.

2. Social movements among persons with disabilities have become very active in recent years in many parts of the world. Some have built on the disease- or condition-specific groups which have lobbied for resources for many decades. Others built on parent groups which were active on behalf of disabled children in the 1940s and 1950s or on successful civil rights activity on behalf of blacks and women in the United States and other western democracies. These movements are challenging the status quo, particularly assumptions about the dependent nature of disability. Demands made by these movements for inclusion, integration and participation come into direct conflict with expensive social

structures put in place to support and maintain persons with disabilities outside the work place, and by default, outside other societal arenas as well. However, the social movements have also created tensions among subgroups of the population identified with disabilities. Social integration and participation for some is limited only by their physical access to the work place or the voting booth. For others with more severe impairments or limitations, full participation is simply not possible in society as it is currently structured, but social movement demands do not recognize this (Ferguson, 1987).

Studies of social movements have by and large ignored the social movements occurring in the disability community. Even within anthologies about various social movements, such as Goldberg's (1991), the disability movement does not show up. The prime exception is one article in Jo Freeman's *Social Movements of the 1960s and 1970s*, which considers the 1977 HEW takeovers (Johnson 1983). There actually have been several social movements, including a disability rights movement, an independent living movement, a movement for deaf rights, a psychiatric survivor's movement, and a parents' movement. These sometimes coalesce and other times focus on their own specific issues (Barnartt and Scotch, 2000 [forthcoming]). Examination of the intricacies of social movements in this community would do much to test the theories and inform the social movement literature that already exists.

3. Demands for inclusion and needs for planning the distribution of scarce resources require that measures of population size need to be available. This has created very real issues about the conceptualization of disability and its subsequent operationalization and measurement. While there is a growing consensus on the theoretical conceptualization of disability, translation of that concept into useful measurement tools is still in its infancy. A highly visible classification scheme (ICIDH) is nearing completion, but is not without its critics (Pfeiffer 1998; Chapireau and Colvez 1998). Its translation into a valid and reliable measurement instrument will require much thought and testing. Numerous other problems associated with data collection (Foster 1996; Mathieowetz in press) loom large as issues to be considered before accurate estimates of this population can be obtained.

4. Both physical and social environments are pivotal issues in understanding all aspects of disability. Some prior research has studied the roles physical barriers play in preventing access to the natural and built environments. Other research has examined the role social barriers, par-

ticularly discrimination, play in affecting integration, access to employment and other important roles. However, both social and physical environments play a key role in *causing* impairments and functional limitations. This latter fact is less well investigated by social science researchers. Violence, which produces traumatic injury, and exposure to toxic chemicals, which can result in cancer or mental deterioration, represent social and physical environmental causes of impairments. Additionally, physical, structural and social barriers contribute to the level of disability experienced following impairment. Social and physical environments need to be examined as much for the risk factors for impairments as for their role in turning functional limitations into disabilities.

5. Many factors contribute to the personal experience of disability and the development of a disability identity. Gender, age at onset, degree of visibility and type of impairment are just a few of the factors which lead to the variety of experiences individuals have living with their impairments. The importance of identity to the person with a disability, and the impact of that identity on psychological and social development, can vary considerably among persons with different types of impairments and from persons without impairments or disabilities. Identity can become an essential component for political activism, but it serves other functions as well. The role of identity in impeding or facilitating integration and inclusion is not known and is a ripe area for research.

6. Rapidly advancing technology is producing assistive devices that are changing the lives of persons with disabilities. Factors which improve access to such devices are not well known, nor is the extent of the impact the devices have on integration and participation. The importance of government programs and insurance in providing these technologies is not well known, and the characteristics of markets needed to stimulate the growth of assistive technology development, production and availability are yet to be studied. Is the economic deck stacked against development or are there untapped markets that business and industry are overlooking? Conceptualizing the role of commerce in dismantling the "dependency" system would contribute to the reassessment of social programs for persons with disabilities.

7. There are also many current ethical and legal issues which are likely to affect persons with disabilities as well as attitudes toward persons with disabilities. Life and death decisions, genetic research and

transplantation all pose serious questions for the value systems of societies. The relationship of the legalization of assisted suicide to the experiences of persons with impairments has not been explored, nor do we know if such changes in laws will pave the way for a eugenics orientation in decision making. The United States and most other societies are currently facing serious ethical decisions related to assisted suicide and other value-laden questions. Decisions in other countries, such as the Netherlands, have already been made and need to be observed closely for adverse effects. Additionally, in the United States interpretations of current law are of concern. Recent Supreme Court decisions appear to be eroding rights of persons with disabilities. These ethical and legal debates provide a rich set of issues to which to apply social science theories and methods.

8. Disability occurs in every known culture. Yet definitions, treatments and experiences of living with a disability are not necessarily the same in any two cultures or even in subcultures within a larger society. Examination of cultural perspectives would contribute to understanding the nature of social barriers and how they are formed. We can learn from other perspectives what can improve the lives of persons with disabilities as well as what practices to avoid. Cross-cultural investigations of economic circumstances, political involvement and value systems would give insights into the subtleties underlying the relationship between culture and disability.

9. Most conceptualization of and research on the issues surrounding disability focus on the transition of an individual from a nondisabled to a disabled status. Very little work has been done on recovery or the transition from disabled to nondisabled. Improvements in assistive technology, prostheses, implants and pharmaceuticals now offer complete or near recovery from some impairments. Certainly they may improve the functioning of the individual. Such changes have rarely been considered and even less frequently researched. However, persons receiving organ implants, artificial hips and other prostheses all experience improvement in functioning at least for a time, yet we know little of the impact of that improvement on the person's access to various roles, their identity, behavior or attitudes.

10. Research in the area of family functioning when there has been a family member with a disability has focused almost exclusively on the dysfunctional elements. For example, there has been much written about care giving to persons with disabilities, but little has been done on

care receiving or other family interactions. This is an important area of concern. Family activity and adaptation, distribution of roles, family formation and disintegration and the nature of extended family involvement for persons with disabilities are among those areas which have yet to be studied extensively.

11. Stratification is another area in which disability has received little attention. Alexander (1976) suggested the use of a status attainment model, similar to that used in studying the socio-economic status of blacks and women, but little has been done in this area (Barnartt 1986). Mostly through the efforts of economists, we know that people with disabilities tend to earn less than people without (Burkhauser and Daly 1994; Johnson and Lambrinos 1985; Baldwin and Johnson 1995). We also know that the incidence of disability varies inversely with social class position, although both causal directions need to be investigated. When the effects of disability are compared to those of race and gender, these effects are usually more serious than those associated with race but follow gender patterns of occupational distribution and income (Barnartt 1997; Barnartt and Altman 1997). Many questions remain about the amount of impact disability has on socio-economic status and the mechanisms by which these impacts operate.

12. Health care is an area which is of increasing political concern. There are many questions related to health care issues for people with disabilities which are likely to be of interest to those studying the health care system. Questions such as whether all people with disabilities use more health care than people without disabilities, whether this is true of only a subgroup or whether it is not true have only begun to be investigated (Altman and Cornelius 1992). Many aspects of the delivery of care to people with disabilities, for example the use of managed care programs, are relevant areas for social science study. Examinations of disability and rehabilitation organizations, such as that done by Albrecht (1992), should be of interest to researchers in several social science fields.

We have begun here to suggest some areas of the disability experience which we feel would be amenable to social science theorizing and research. The list is not meant to be exhaustive but to stimulate thought and discussion. There are also established concepts, topics and perspectives within the social science disciplines which suggest further fruitful ways these disciplines might approach disability. We consider these below.

DISCIPLINARY PERSPECTIVES

Sociological Perspectives

Disability most commonly is discussed within the context of medical sociology, although it is—ironically—not often included in medical sociology textbooks (Barnartt 1990, 1995). Medical sociology, which locates itself both within and outside medicine, has primarily approached disability as a medical condition that needs to be cured or rehabilitated. Until very recently medical sociology in general has missed the bigger picture of individuals with functional limitations in a social environment. In addition, the discipline as a whole has considered disability as a phenomenon which can be analyzed only from a medical sociology point of view and has rarely been aware that many of the core concepts and theories within sociology also apply to disability. For example, many deaf people feel strongly that there is a deaf culture, and many people with other types of disabilities feel that there is a disability culture. Are these true cultures? Or are they subcultures or countercultures? These are questions that could sustain a cultural sociology investigation.

Sociologists could usefully apply the concepts of role and status to disability. "Person with a disability" is a role, but its similarities to, and differences from, other roles have not yet been delineated, nor has its influence on other roles such as wife and mother been fully explored. To this point, sociologists have not considered the question of whether or not disability is a master status, and, if so, how it compares to other master statuses such as race and gender. Barnartt and Scotch (2000) make a start at such analyses.

Other specialty areas in sociology including the areas of race, ethnicity or minority relations, demography and social psychology could usefully examine disability to a greater extent than has been accomplished to date. Comparisons of traditionally defined racial, ethnic, and minority groups with people with disabilities on issues such as group ties, cultural transmission, power relationships and prejudice and discrimination might be fruitful avenues of exploration. Sociologists considering the topic of socialization could discuss the effects of having a disability on early socialization of children, including its affect on parent-child interaction and ways in which parents and society create dependency in children with disabilities. The effects of adult-onset disabilities as an example of secondary socialization or examination of disabilities through the life cycle, especially the relationship between late-onset disabilities and aging, would be useful additions to the literature.

Demographic and epidemiological issues have been addressed to some extent (Nagi 1969, 1976; Verbrugge et al. 1991; LaPlante and Carlson 1996; McNeil 1993) but there is room for further understanding about distributions of potentially disabling conditions, or incidence or prevalence of impairments by age, gender and race in the United States and other countries. Demographers have not yet researched traditional demographic variables such as fertility, life cycle transitions and morbidity and mortality rates for persons with impairments.

Sociologists who specialize in deviance have been, along with medical sociologists, those most likely to consider disability, although they have mostly considered it to be a prime example of stigma-producing condition (e.g. Goffman 1963) or an example of how the labeling process works (Haber and Smith 1971). However, criminologists have not paid much attention to disability, except in considerations of victimization. Research into the experiences of people with impairments as they proceed through the criminal justice system, and the impact of criminal actions (such as shootings, spousal abuse and rape) as causal factors in the production of disability among some population subgroups would be natural extensions of criminology perspectives into the disability arena.

Other social sciences also have many perspectives that they can bring to addressing the issues that surround disability in modern society. As sociologists, we do not have the in-depth familiarity with these disciplines that we have with our own, but the following highlights contributions from those other disciplines that we feel are possible and necessary.

Psychological Perspectives

Traditionally, psychological approaches to disability have addressed the emotional and personality characteristics of the individual who is impaired. The focus can be on the cognitive, personality development or adaptive mechanisms of the individual during and after the onset of impairment and subsequent disability. From a psychological perspective, disability is a dynamic process rather than a static state. Questions that psychologists pose frequently attempt to capture that dynamic process by studying the coping mechanisms individuals develop or by examining the motivational basis for individual changes. In many instances of psychological research, coping, or the changed social behavior, is the dependent variable and not the starting point of research which it could be. Frequently, the units of analysis are defined by the pathology or disease state. From this perspective any number of physical diseases or dysfunction can be used to identify the "potentially disabled" including heart disease, arthritis or even obesity. The nature of the environmental factors that contribute to the development of disability are not often considered in psychological research as it is now formulated. It would be useful for psychologists to explore the interactive aspects of disability from a broader perspective. The issue of identity and identity formation when one has a disability is also a natural area for psychologists.

Social Psychology Perspectives

Social psychology can be an offshoot either of sociology or psychology. Sociological social psychology is most likely to focus on dyadic or small group interactions, attitudes and attitude changes, and relationships among roles, identities, the self and others' perception of self. There are many topics within sociological social psychology which are amenable to disability-related inquiries. One might examine ways in which impairments change people's patterns of interaction, including the effects of stigmatized conditions on interaction patterns, of differences in interaction patterns depending upon whether the disability is visible or invisible and of impression management by people with impairments. The attitude literature has examined attitudes toward disability in general as well as attitudes toward specific disabilities.

Psychological social psychology includes interest in motivation, attribution processes, interpersonal attraction, cognitive processes and learning theory. Some of the topics on which sociological and psychological perspectives overlap include attitude formation and the relationships between attitudes and behavior. The conception of identity and self-evaluation for a person with a disability has been explored to a certain extent by psychological social psychologists, but issues of self-estrangement and the integration of multiple roles with a disabled identity are ripe areas for further investigation.

With the new paradigms or models of disability that focus more clearly on the social and physical environment, the place of social roles and the characteristics of social roles in the lives of persons with disabilities need to be reexamined and elaborated. Social psychology, in particular, brings a set of theories and a literature that is very relevant to the experience of a person with a disability in today's society.

Anthropological Perspectives

Anthropology has traditionally focused upon more "primitive" cultures and the members of those cultures, although, more recently, anthropological concepts and methods have been brought to bear upon people and groups within their own societies and times. Cultures are studied either individually (ethnography) or collectively (ethnology). Kasnitz and Shuttleworth (1999) note that the ethnological approach to otherness, to difference, to "not of us," as a topic of study is a uniquely compelling aspect of anthropology which makes it a natural discipline to engage in disability studies.

The concept which underlies the work of those who practice cultural anthropology is that of culture, although it is a concept on whose definition anthropologists do not agree (Griswold 1994). Possible application of the concept of culture to people with disabilities, including people who are deaf, is a subject of some controversy within those communities. The debate is one into which anthropological definitions might usefully be injected. Additionally, differences between views of disability in various cultures have just begun to be analyzed (Ingstad and Whyte 1995), and generalities about views of disability within all known cultures have begun to be suggested (Groce and Scheer 1990).

Anthropologists analyze cultural conceptions of all aspects of social order, including families, kinship patterns, economic systems, religious systems, health care systems, military systems, political systems and all forms of art, including music, dance, visual representations and drama. Almost none of these analyses have paid attention to disability in these realms, either within the United States or in other countries.

Historical Perspectives

History is an attempt to reconstruct the past. However, as with blacks and women, historians have been slow to recognize the participation of people with disabilities in events over time. Very little history acknowledges contributions of persons with disabilities nor has history been examined from the point of view of persons with disabilities. (Some notable exceptions include Trent 1994; Ferguson 1994; Longmore 1997). As J. Hirsch (1998, p. 265) notes:

> For too long, historians have failed to see that they have constructed a master narrative that excludes and silences disabled individuals—a narrative that unreflectively privileges the history of the able-bodied as the history of humanity.

Partly this was true because the data of history are, in the main, produced by those with power rather than those without. The latter group indubitably includes many people with disabilities. However, some historical studies have used organizational or institutional documentation to good effect to discuss the social creation of disability categories and policies (Trent 1994).

The situation in the field of history has changed recently, as more historians have turned their attention to people with various types of impairments. In the United States, some have focused on the ways in which our society created categories of people with disabilities and differentiated among them (e.g., Liachowitz 1988), while others focused on the history of policies for people with disabilities in general (e.g., Berkowitz 1987) or of policies for people with specific types of impairments (Ferguson 1994). Historians have examined historical situations for people with mobility impairments (e.g., Longmore 1997), visual impairments (Matson 1990) and hearing impairments (e.g., Van Cleve and Crouch 1989). Still others have examined the histories of people with disabilities in other countries (Mottez 1993) or time periods (Winzer 1997). There are still many aspects of the history of people with many kinds of disabilities or impairments about which we know nothing.

Political Science Perspectives

Political science is concerned with all types of political behavior, from campaigning to voting to lobbying to conducting revolutions. It is concerned with the behavior of politicians and ruling elites as well as those who are neither. It is concerned with all political actors: people, organizations, governments and countries. It is concerned with the results of political behavior, including policies, laws and regulations at all levels of government.

Since disability policy makes up a recognizable amount of national policy in the United States and many developed countries, the expectation is that political science would demonstrate more integrated research in this area. However, that is not the case. In addition, Hahn (1993) has pointed out that the study of disability issues has led to the uncovering of anomalies in several of the dominant paradigms of political science. Is this a deterrent or a challenge to those who work within the political science field? From another viewpoint, Fox (1994) found that research on disability policy had achieved a great deal in a short time, but much remains to be done. He particularly noted that the work on research policy has been "relentlessly reformist" and insular (Fox 1994, p. 166) but urged that doing research that cut across disabling conditions and using an objective and neutral perspective of analysis would go far to help the field realize its potential. Political science has been particularly challenged to develop a research agenda in disability policy that can contribute to the development as well as the evaluation of the processes that are taking place relative to disability in societies today.

Economics Perspectives

One economic perspective is based on three basic observations about the world and the way it works (Fuchs 1974). First, resources are not unlimited; they are finite and generally cannot be expanded quickly or easily. Second, resources have alternative uses, so that multiple groups are seeking to use available resources at the same time. Finally, people do have different wants and needs and there is significant variation in the relative importance that they attach to them. While this perspective was written as an introduction to the study of health economics, it is particularly relevant for economics research in the area of disability.

In the short review of disability research presented above, we indicated some important contributions on employment, wages and disability benefits that have been made by economists working in this area. However, there are many other research avenues relevant to persons with disabilities that economists might pursue. The work by Baldwin and Johnson (1995) on discrimination toward persons with disabilities would benefit from further elaboration, particularly as a follow-up to the ADA. The role of markets in providing all the necessary health care to persons with disabilities has yet to be studied completely and the relationship of the labor market to employment for persons with disabilities has been questioned (Yelin 1992) and needs follow-up investigation. The changing nature of long-term care, home care, and personal assistant services also provide an interesting and essential avenue of research for interested economists. We hope that we can publish such important research in future volumes.

We have suggested just a few examples of ways in which disability might usefully be examined using some of the core concepts within sociology, psychology, anthropology, social psychology, history, political science and economics. We know our readers can think of many more and hope that this sampling will generate exciting new ideas to help advance the field.

Measurement and Methodologies

The social science disciplines mentioned above, as well as those not included in this brief overview, provide a rich source of measurement and methodologies with which to approach research on disability. Quantitative and qualitative methodologies are well represented in their repertoires and provide useful tools to gain insight into the experience, structure, function and other aspects of disability. We feel that measurement tools and methodologies are so important to the study of disability and particularly to the expansion of that study, that we will be devoting the next volume of *Research in Social Science and Disability* to that topic. In that volume, we will continue this overview with specific attention to the variety of methodologies available for use in the social sciences, highlighting the benefits and limitations of those methods with particular emphasis on the issue of definition and measurement of the central concept of disability. Calls for papers for that volume, which will be the second in this series, have already been published.

In This Volume

This first volume of *Research in Social Science and Disability* is a generic one with a wide variety of topics, methodologies and perspectives. The papers we present in this volume range from a new qualitative methodology for studying health services for persons with disabilities to new historical material which is being published here for the first time. Many of the papers are moving into unexplored areas in the study of disability and others are adding strong evidence to established areas. We are excited to share these advances with you.

Chris Borthwick, in the paper *Idiot into Ape*, presents an analysis of material contained in letters from 1873 between Charles Darwin and Dr. L. Down, the physician who identified Down Syndrome. Few probably know that Darwin and Down were contemporaries and this evidence of a personal relationship helps us understand how the theory of evolution got translated and applied to persons with mental retardation, particularly those with Down Syndrome. Borthwick follows the lines of reasoning to present day and the application of training methods that have been involved with trying to teach primates to communicate with persons with severe cognitive impairments. Interesting questions about appropriate techniques to teach humans with cognitive limitations are raised as well as questions about how conceptions of persons with mental retardation as throwbacks influence the development of stigma for these individuals.

In the second paper, Beth Haller presents a very interesting analysis of the media coverage of the ADA. Using content analysis and elaborating the basic media model developed by Clogston (1990), Haller reviews news and feature articles of nine national newspapers and three news magazines for the period from 1988 to 1993. She also includes an analysis of photographs which accompany some of the stories. This very timely research gives us a good understanding of how the media present the issues associated with the ADA and what sources they see as relevant to interpreting the information. While she found that architectural access was a primary issue covered by the media, general discrimination and access to jobs also received reasonable coverage. However, she found misrepresentation in media coverage that is seen as contributing to the backlash against the ADA. In all this is a valuable investigation of media interpretation and presentation of disability issues that should go a long way toward generating more and continued media monitoring on these issues.

A distinguished sociologist, Judith Lorber, has contributed a thoughtful analysis of gender issues associated with disability. She notes that both men and women have benefited from the successes of the disability rights movement, but that special needs of women with disabilities have not been fully addressed by these advances. She discusses the gender differences in the roles and statuses of people with disabilities and argues that there are numerous contradictions and status problems in beliefs about the characteristics of women and men with disabilities versus general gender expectations. She suggests that these contradictions create constraints for both men and women with disabilities and discusses examples such as the expectation that care givers will be women. Improved independence and integration will benefit from this examination of the overlay of gender roles on disability status.

A particularly thoughtful and challenging perspective on labor market factors which impact on persons with disabilities has been contributed to this collection by Edward Yelin and Laura Trupin. Using data from the Current Population Survey (CPS), the authors analyze the factors affecting the probability that persons with disabilities will be able to enter new jobs or maintain the ones they hold. Their findings are particularly disturbing in that persons with disabilities are about 30 percent as likely to be employed at any one time as persons without disabilities and, if unemployed, only about one fifth as likely to enter jobs. However, they also found that demographic and work-characteristic differences between persons with and without disabilities account for a substantial portion of the gap in employment rates but none of the difference in the ability to gain entry to new jobs. It should be noted, however, that the measure of disability in the CPS is limited to a question that asks respondents whether they have health limitations that prevent work or limit the kind or amount of work that can be done. Because of that, the analysis reflects only a portion of the population that current theory would identify as having a disability.

Marie Campbell has contributed a fascinating paper that describes a form of institutional ethnography undergoing development that could be used to understand the experience of persons with disabilities in health care settings. She discusses the concept of performativity, which she defines as administrative practices enacted by people and arranged in conjunction with technological processes. She uses this concept as the basis for her investigation of a specific case of public health care. Dr. Campbell's purpose is to elaborate the methodology for deriving the institutional practices that will then be observed in a study of the organization in question. As usual with ethnographic material, the paper provides a very rich data source for generating the imbedded institutional norms and behaviors that are sought. In the current understanding of the importance of environment to the creation of disability, this methodology that helps to get at the context in which health care is provided can prove to be very useful for future research on many disability related environments.

From a much different health perspective, Lee Cornelius has contributed a paper that examines hospitalizations for violent/intentional injuries. Violence associated with intentional injury has become a public health problem for the entire population. It is of concern also to those who are interested in disability for two important reasons: (1) intentional injuries can result in impairments and subsequent disability; and (2) persons with disabilities may be more vulnerable to violence and intentional injury because of their disability status. Using data from a sample of 800 hospitals in 11 states, Dr. Cornelius investigates the characteristics of persons hospitalized for intentional injuries. His findings that persons with disability-related conditions are more likely to be hospitalized for such injuries substantiates much of the anecdotal data about the vulnerability of persons with disabilities to such threats. His work helps define new areas of research that are needed in order to understand this finding and to prevent such violence among this more vulnerable population.

Shields, Schriner, Schriner and Ochs' contribution to this volume addresses a new area in disability research, mentioned above in our review of new research opportunities. They examine the voting behavior of persons with disabilities in recent national elections. Reviewing the literature on voting behavior among persons with disabilities, they find that the literature suggests that this population is much less likely to take part in the electoral process. Their review identifies several problems associated with this lack of participation and also points to several limitations in the data available with which to analyze these kinds of questions. Of particular interest are the questions that they raise for future study of the political activity of this population.

Finally, a case study of the implementation of the ADA authored by Sanjoy Mazumdar and Gilbert Geis brings another new line of research in public policy. Drs. Mazumdar and Geis have done an excellent in-depth analysis of two court cases associated with the line of sight available to persons in wheelchairs in newly built stadiums. They highlight the vagueness and ambiguity that the implementation of law can create and urge the necessity of clarification in the formulation of the law as well as persistence in the courts to fully interpret laws already on the books. An important point made in the analysis is the political nature of the legal process involved in court interpretations of the law and the importance of this in pursuing clarification of laws that apply to persons with disabilities. We are quite excited about the potential of this volume series and the papers that make up this first volume. We hope this material is useful to you, the reader. We look forward to increasing submissions to these volumes and opening new areas of research in social science and disability.

ACKNOWLEDGMENTS

We would like to thank Paul Higgins and Gary Albrecht for suggesting, encouraging and supporting the development of this series. We would also like to thank the Editorial Board and the other additional reviewers for their thoughtful and helpful reviews of the manuscripts submitted for this first issue.

NOTE

1. We must note, however that this conceptualization is not without its problems or detractors. Legal definitions are not based on conceptual models and often are confused with scholarly definitions by popular or lay usage of terminology. Legal definitions are used to determine receipt of disability benefits and can only be used as scholarly definitions for purpose related to receipt of those benefits or examinations of policy definitions.

REFERENCES

Abberley, P. 1987. "The Concept of Oppression and the Development of a Social Theory of Disability." *Disability, Handicap and Society* 2: 5-12.
Albrecht, G. 1976. *The Sociology of Physical Disability and Rehabilitation*. Pittsburgh, PA: University of Pittsburgh Press.
_____. 1992. *The Disability Business*. London: Sage.
Albrecht, G., and J. Levy. 1981. "Constructing Disabilities as Social Problems." In *Cross National Rehabilitation Policies: A Sociological Perspective*, edited by G. Albrecht. Beverly Hills and London: Sage Publications.
Alexander, K.L. 1976. "Disability and Stratification Processes." In *The Sociology of Physical Disability and Rehabilitation,* edited by G. Albrecht. Pittsburgh, PA: University of Pittsburgh Press.
Altman, B.M. 1981. "Studies of Attitudes Toward the Handicapped: The Need for a New Direction." *Social Problems* 28(3): 321-337.
_____. 1984. *Examination of the Effects of Individual, Primary and Secondary Resources on the Outcomes of Impairment*. Doctoral Dissertation, University of Maryland, College Park, MD.

_____. 1993. "Definitions of Disability and their Measurement and Operationalization in Survey Data." Pp. 219-224 in *Proceedings of the 1993 Public Health Conference on Records and Statistics.* Washington, DC: National Center for Health Statistics, DHHS.

_____. 1997. "Does Access to Acute Medical Care Imply Access to Preventive Care?" *Journal of Disability Policy Studies* 8(1&2): 99-128.

Altman, B.M., and L. Cornelius. 1992. "High Users or Neglected Minority? Access to Health Care Among Working-age Persons with Disabilities." Presented at the annual meetings of the Society for Disability Studies, Rockville, MD.

Badley, E.M. 1987. "The ICIDH: Format, Application in Different Settings, and Distinction Between Disability and Handicap." *International Disability Studies* 9(3): 122-125.

Baldwin, M.L., and W. Johnson. 1995. "Labor Market Discrimination Against Women with Disabilities," *Industrial Relations* 34(4): 55-77.

Barnartt, S.N. 1986. "Disability as a Socioeconomic Variable: Predicting Deaf Worker's Incomes." Presented at the annual meeting of the American Sociological Association, New York.

_____. 1990. "A Review of Medical Sociology Textbooks." *Teaching Sociology* 18 (July): 372-376.

_____. 1995. "Medical Sociology Textbooks Reconsidered." *Teaching Sociology* 23(1): 69-74.

_____. 1997. "Gender Differences in Changes over Time: Educations and Occupations of Adults with Hearing Losses 1972-1991." *Journal of Disability Policy Studies* 8(1).

Barnartt, S.N., and B.M. Altman. 1997. "Predictors of Wages: Comparisons by Gender and Type of Impairment." *Journal of Disability Policy Studies* 8(1): 51-74.

Barnartt, S.N., and R. Scotch. forthcoming. *Contentious Political Actions in the Disability Community: 1970-1999.* Washington, DC: Gallaudel University Press.

Barry, J., G. Dunteman, and M. Webb. 1968. "Personality and Motivation in Rehabilitation." *Journal of Counseling Psychology*

Ben-Sira, Z. 1981. "The Structure of Readjustment of the Disabled: An Additional Perspective on Rehabilitation." *Social Science and Medicine* 15A: 565-581.

_____. 1983. "Loss, Stress and Readjustment: The Structure of Coping and Bereavement and Disability." *Social Science and Medicine* 21: 1619-1632.

Berkowitz, E. D. 1987. *Disabled Policy: America's Programs for the Handicapped.* New York: Cambridge University Press.

Berkowitz, M. 1973. "Workmen's Compensation Income Benefits: Their Adequacy and Equity." Pp. 189-268 in *Principles of Workmen's Compensation,* edited by M. Berkowitz. Washington, DC: U.S. Government Printing Office.

Berkowitz, M., J.F. Burton, and W. Vroman. 1979. *An Evaluation of State Level Human Resource Delivery Programs: Disability Compensation.* Washington, DC: National Science Foundation.

Berkowitz, M., W.G. Johnson, and E.H. Murphy. 1971. *Measuring the Effects of Disability on Work and Transfer Payments: A Multivariate Analysis.* Report for the Social Security Administration, Bureau of Economic Research, Rutgers University.

Bogdan, R., and D. Biklen. 1977. "Handicapism." *Social Policy* 7: 14-19.

Brandt, E.N., and A.M. Pope (eds.). 1997. *Enabling America*, Washington, DC: National Academy Press.
Brown J., and M. Rawlinson. 1976. "The Morale of Patients Following Open-heart Surgery." *Journal of Health and Social Behavior* 17: 135-145.
Brown, S. 1990. "Conceptualizing and Defining Disability." In *Disability in the United States: A Portrait from National Data*, edited by S. Thompson-Hoffman and I.F. Storck. New York: Springer Publishing.
Burkhauser, R.V., and M.C. Daly. 1994. "The Economic Consequences of Disability." *Journal of Disability Policy Studies* 5(1): 25-52.
Campbell, J., and M. Oliver. 1996. *Disability Politics: Understanding Our Past, Changing Our Future*. London: Routledge.
Cassem, N., and T. Hackett. 1973. "Psychological Rehabilitation of Myocardial Infarction Patients in the Acute Phase." *Heart and Lung* 2(3): 382-388.
Chapireau, F., and A. Colvez. 1998. "Social Disadvantage in the International Classification of Impairments, Disabilities, and Handicap." *Social Science and Medicine* 47(1): 59-66.
Christiansen, J.B., and S.N. Barnartt. 1987. "The Silent Minority: The Socioeconomic Status of Deaf People." Pp. 171-196 in *Understanding Deafness Socially*, edited by P. Higgins and J. Nash. Springfield, IL: Charles C. Thomas.
Clagston, J.S. 1990. *Disability Coverage in 16 Newspapers*. Louisville, KY: Advocado Press.
Cohn, N. 1970. "Understanding the Process of Adjustment to Disability." *Journal of Rehabilitation* 10: 16-20.
Conley, R.W. 1965. *The Economics of Vocational Rehabilitation*. Baltimore, MD: Johns Hopkins University Press.
_____. 1973. *The Economics of Mental Retardation*. Baltimore, MD/London: Johns Hopkins University Press.
Crammette, A. 1968. *Deaf Persons in Professional Employment*. Springfield, IL: Charles C. Thomas.
Croog, S., and S. Levine. 1977. *The Heart Patient Recovers: Social and Psychological Aspects*. New York: Human Services Press.
Davis, F. 1963. *Passage Through Crisis*. Indianapolis, IN: Bobbs-Merrill Company.
DeJong, G. 1979. "Independent Living: From Social Movement to Analytic Paradigm." *Archives of Physical Medicine and Rehabilitation* 60: 435-440.
DeJong, G., and T. Wenker. 1983. "Attendant Care as a Prototype Independent Living Service." *Caring* 2: 26-30.
Ferguson, P.M. 1987. "The Social Construction of Mental Retardation." *Social Policy* 18(1): 51-56.
_____. 1994. *Abandoned to Their Fate: Social Policy and Practice Toward Severely Retarded People in America 1820-1920*. Philadelphia, PA: Temple University Press.
Finlayson, A., and J. McEwen. 1977. *Coronary Heart Disease and Patterns of Living*. New York: Prodist.
Foster, S. 1996. "Doing Research in Deafness: Some Considerations and Strategies." Pp. 3-20 in *Understanding Deafness Socially* (2nd edition), edited by P. Higgins and J. Nash. Springfield, IL: Charles C. Thomas.

Fox, D. 1994. "The Future of Disability Policy as a Field of Research. *Policy Studies Journal* 22(1): 161-167.
Fuchs, V. 1974. *Who Shall Live? Health Economics and Social Choice.* New York: Basic Books.
Gliedman, J., and W. Roth. 1980. *The Unexpected Minority.* New York: Harcourt Brace Jovanovich.
Goffman, E. 1963. *Stigma.* Englewood Cliffs, NJ: Prentice Hall, Inc.
Goldberg, R.A. 1991. *Grassroots Resistance: Social Movements in Twentieth Century America.* Prospect Heights, IL: Waveland Press.
Griswold, W. 1994. *Cultures and Societies in a Changing World.* Thousand Oaks, CA: Pine Forge Press.
Groce, N. 1985. *Everyone Here Spoke Sign Language: Hereditary Deafness on Martha's Vineyard.* Cambridge, MA: Harvard University Press.
Groce, N., and J. Scheer. 1990. Introduction. *Social Science and Medicine* 30(8): v-vi.
Haber, L., and R.T. Smith. 1971. "Disability and Deviance: Normative Adaptations of Role Behavior." *American Sociological Review* 36: 87-97.
Hahn, H. 1982. "Disability and Rehabilitation Policy: Is Paternalistic Neglect Really Benign?" *Public Administration Review* 42(4): 385-389.
_____. 1983. "Paternalism and Public Policy." *Society* March/April: 36-46.
_____. 1985. "Toward a Politics of Disability: Definition, Disciplines and Politics." *The Social Science Journal* 22(4): 87-105.
_____. 1993. "The Potential Impact of Disability Studies on Political Science (as well as Vice-Versa)." *Policy Studies Journal* 21(4): 740-751.
Haveman, R.H., V. Halberstadt, and R.V. Burkhauser. 1984. *Public Policy Toward Disabled Worker: Cross-National Analyses of Economic Impacts.* Ithaca, NY: Cornell University Press.
Higgins, P.C. 1992. *Making Disability: Exploring the Social Transformations of Human Variation.* Springfield, IL: Charles C. Thomas.
Hirsch, J. 1998. "History and a Story of Polio." *Disability Studies Quarterly* 18(4): 264-266.
Ing, C.D., and B.P. Tewey. 1994. *Summary of Data on Children and Youth with Disabilities.* U.S. Department of Education, National Institute on Disability and Rehabilitation Research, Washington, DC.
Ingstad, B., and S.R. Whyte. 1995. *Disability and Culture.* Berkeley, CA: University of California Press.
Johnson, R.A. 1983. "Mobilizing the Disabled." Pp. 82-97 in *Social Movements of the 60's and 70's,* edited by Jo Freeman. New York: Longman.
Johnson, W.G. 1979. "Disability, Income Support, and Social Insurance." In *Disability Policies and Government Programs,* edited by E.D. Berkowitz. New York: Praeger.
Johnson, W.G., and J. Lambrinos. 1985. "Wage Discrimination Against Handicapped Men and Women." *Journal of Human Resources* 20(2): 264-277.
Kasnitz, D., and R.P. Shuttleworth. 1999. "Engaging Anthropology in Disability Studies." Presented at the Society for Disability Studies Annual Meeting, Washington, DC.
Kaye, S., M.P. Laplante, D. Carlson, and B.L. Wenger. 1996. "Trends in Disability Rates in the United States, 1970-1994." *Disability Statistics Abstract, Number*

17. National Institute of Disability and Rehabilitation Research, Washington, DC.

Kelman, H., J. Miller, and M. Lowenthal. 1964. "Post-hospital Adaptation of a Chronically Ill and Disabled Rehabilitation Population." *Journal of Health and Social Behavior* 5: 108-114.

LaPlante, M.P. 1991a. "Disability in Basic Life Activities Across the Life Span." *Disability Statistics Report 1.* National Institute on Disability and Rehabilitation Research, Washington, DC.

LaPlante, M.P. 1991b. "Disability Risks of Chronic Illnesses and Impairments." *Disability Statistics Report 2.* National Institute on Disability and Rehabilitation Research, Washington, DC.

_____. 1993. "State Estimates of Disability in America." *Disability Statistics Report 3.* National Institute on Disability and Rehabilitation Research, Washington, DC.

LaPlante, M.P., and D. Carlson. 1996. "Disability in the United States: Prevalence and Causes, 1992." *Disability Statistics Report 7.* National Institute on Disability and Rehabilitation Research, Washington, DC.

Liachowitz, C. 1988. *Disability as a Social Construct: Legislative Roots.* Philadelphia, PA: University of Pennsylvania Press.

Litman, T.J. 1966. "The Family and Physical Rehabilitation." *Journal of Chronic Disability* 19: 211-220.

Longmore, P. 1997. "Political Movements of People with Disabilities: The League of the Physically Handicapped, 1935-1938." *Disability Studies Quarterly* 17(2): 94-98.

Ludwig, E.G., and J. Collette. 1970. "Dependency, Social Isolation and Mental Health in a Disabled Population." *Social Psychiatry* 5: 92-95.

Mathiowetz, N.A. (in press). "Methodological Issues in the Measurement of Work Disability." In *Workshop on Survey Measurement of Work Disability,* edited by N. Mathiowetz and G. Wunderlich. Washington, DC: National Academy Press.

Matson, F. 1990. *Walking Alone and Marching Together: A History of the Organized Blind Movement in the United States, 1940-1990.* Baltimore, MD: The National Federation of the Blind.

McNeil, J. 1993. *Americans with Disabilities: 1991-92: Data from the Survey of Income and Program Participation.* Pp. 33-70 in Current Population Report. U.S. Department of Commerce, Bureau of the Census: Washington, DC.

Meer, C.W. 1979. *Social Security Disability Insurance: The Problems of Unexpected Growth.* Washington, DC: American Enterprise Institute.

Monaghen, P. 1998. "Pioneering Field of Disability Studies Challenges Established Approaches and Attitudes." *Chronicle of Higher Education* (Jan 23): 15-16.

Mottez, B. 1993. "The Deaf-Mute Banquets and the Birth of the Deaf Movement." Pp. 27-40 in *Deaf History Unveiled: Interpretations from the New Scholarship,* edited by J.V. Van Cleve. Washington, DC: Gallaudet University Press.

Nagi, S.Z. 1965. "Some Conceptual Issues in Disability and Rehabilitation." In *Sociology and Rehabilitation,* edited by M.B. Sussman. Washington, DC: American Sociological Association

_____. 1969. *Disability and Rehabilitation: Legal, Clinical and Self Concepts and Measurement.* Columbus, OH: Ohio University Press.

_____. 1976. "An Epidemiology of Disability Among Adults in the United States." *Milbank Memorial Fund Quarterly* 54: 439-467.
Neath, J. 1997. "Social Causes of Impairment, Disability and Abuse: A Feminist Perspective." *Journal of Disability Policy Studies* 8(1&2): 195-230.
New, P., A. Ruscio, R.P. Priest, D. Petritsi, and L. George. 1968. "The Support Structure of Heart and Stroke Patients." *Social Science and Medicine* 2: 185-200.
Nosek, M.A., C.A. Howland, and M.E.Young. 1997. "Abuse of Women with Disabilities." *Journal of Disability Policy Studies* 8(1&2): 157-175.
Oliver, M. 1990. *The Politics of Disablement*. Basingstoke: Macmillan.
O'Neill D. 1976. *Discrimination Against Handicapped Persons: The Costs, Benefits, and Inflationary Impact of Implementing Section 504 of the Rehabilitation Act of 1973 Covering Recipients of HEW Financial Assistance*. Report to the Office for Civil Rights. Arlington, VA.: Public Research Institute.
Peck, C. 1983. "Employment Problems of the Handicapped: Would Title VII Remedies Be Appropriate and Effective." *Journal of Law Reform* 16(2): 343-385.
Pfeiffer, D. 1998. "The ICIDH and the Need for its Revision." *Disability and Society* 13(4): 503-523.
Petersen, Y. 1979. "The Impact of Physical Disability on Marital Adjustment: A Literature Review." *Family Coordinator*: 47-51.
Pope, A., and A. Tarlov. (eds.). 1991. *Disability in America*. Washington, DC: National Academy Press.
Richardson, S.A. 1970. "Age and Sex Differences in Values Toward Physical Handicaps." *Journal of Health and Social Behavior* 11: 207-214.
_____. 1971. "Children's Values and Friendships: A Study of Physical Disability." *Journal of Health and Social Behavior* 12(3): 253-258.
Richardson, S.A., and J. Royce. 1968. "Race and Physical Handicap in Children's Preference for Other Children." *Child Development* 39: 467-480.
Safilios-Rothchild, C. 1970. *The Sociology and Social Psychology of Disability and Rehabilitation*. New York: Random House.
Schroedel, J. (ed.). 1978. *Attitudes Toward Persons with Disabilities: A Compendium of Related Literature*. Albertson, NY: Human Resources Center.
Scheffler, R., and G. Iden. 1974. "The Effect of Disability on Labor Supply." *Industrial and Labor Relations Review* 28: 122-132.
Schriner, K.F., T. Shields, and K. Schriner. 1998. "The Effect of Gender and Race on the Political Participation of People with Disabilities in the 1994 Mid-term Election." *Journal of Disability Policy Studies* 9: 53-76.
Shakespeare, T. 1994. "Cultural Representation of Disabled People: Dustbins for Disavowal." *Disability and Society* 9(3): 283-299.
Shears, L.M., and C.J. Jensema. 1969. "Social Acceptability of Anomalous Persons." *Exceptional Children* 36(2): 91-96.
Shields, T., K.F. Schriner, and K. Schriner. 1998. "The Disability Voice in American Politics: Political Participation of People with Disabilities in the 1994 Election. *Journal of Disability Policy Studies* 9: 33-52.
Siller, J., and A. Chipman. 1964. "Factorial Structure and Correlates of the Attitudes Toward Disabled Persons Scale." *Educational and Psychological Measurement* 24(4): 831-840.

Shontz, F. 1975. *The Psychological Aspects of Physical Illness and Disability*. New York: Macmillan.
Slade, F.P. 1984. "Older Men, Disability Insurance and the Incentive to Work." *Industrial Relations* 23(2): 260-277.
Smith, R.T. 1981. "The Role of Social Resources in Cardiac Rehabilitation." In *Physical Conditioning and Cardiovascular Rehabilitation*, edited by L.S. Cohen et al. New York: John Wiley and Sons.
_____. 1979. "Disability and the Recovery Process: Role of Social Networks." In *Patients, Physicians and Illness* 3rd edition, edited by E.G. Jaco. New York: Free Press.
Starkey, P.D. 1968. "Sick-role Retention as a Factor in Non-rehabilitation." *Journal of Counseling Psychology* 15: 75-79.
Stroman, D.F. 1982. *The Awakening Minorities*, Washington, DC: University Press of America.
Swain, J., V. Finkelstein, S. French, and M. Oliver. 1993. *Disabling Barriers—Enabling Environments*. Newbury Park, CA: Sage Publications.
Swisher, I. 1973. "The Disabled and the Decline in Men's Labor Force Participation." *Monthly Labor Review* 96: 53.
Tolsdorf, C.C. 1976. "Social Networks, Support and Coping: An Exploratory Study." *Family Process* 15: 407-409.
Trent, J. 1994. *Inventing the Feeble Mind: A History of Mental Retardation in the U.S.* Berkeley, CA: University of California Press.
Van Cleve, J.V., and B.A. Crouch. 1989. *A Place of their Own: Creating the Deaf Community in America*. Washington, DC: Gallaudet University Press.
Verbrugge, L. 1990. "The Iceberg of Disability." In *The Legacy of Longevity, Health and Health Care in Later Life*, edited by S. Stahl. Newbury Park, PA: Sage Publications.
Verbrugge, L.M., and A.M. Jette. 1994. "The Disablement Process." *Social Science and Medicine* 38(1): 1-14.
Verbrugge, L.M., J.M. Lepkowski, and L.L. Konkol. 1991. "Levels of Disability Among U.S. Adults with Arthritis." *Journal of Gerontology* 46(2): 871-883.
Wan, T.H.T. 1974. "Correlates and Consequences of Severe Disabilities." *Journal of Occupational Medicine* 16(4): 234-244.
Winzer, M.A. 1997. "Disability and Society Before the Eighteenth Century." Pp. 75-109 in *The Disability Studies Reader*, edited by L.J. Davies. New York: Routledge.
Wood, P.H.N. 1980. "Appreciating the Consequences of Disease: The International Classification of Impairments, Disabilities, and Handicaps." *WHO Chronicle* 34: 376-380.
Worrall, J. 1978. "A Benefit-cost Analysis of the Vocational Rehabilitation Program." *Journal of Human Resources* 13(2): 285-298.
Wright, B. 1960. *Physical Disability—A Psychological Approach*. New York: Harper and Row.
Yelin, E. 1980. "Work Disability in Rheumatoid Arthritis: Effects of Disease, Social and Work Factors." *Annals of Internal Medicine* 93: 551-556.
Yelin, E.H. 1992. *Disability and the Displaced Worker*, New Brunswick, NJ: Rutgers University Press.

Yuker, H.E., J.R. Block, and W. Campbell. 1960. *A Scale to Measure Attitudes Toward Disabled Persons*. Albertson, NY: Human Resources Foundation.

Yuker, H.E., J.R. Block, and J.H. Young. 1966. "The Measurement of Attitudes Toward Disabled Persons." *Human Resources Study No. 7.* Albertson, NY: Human Resources Center.

IDIOT INTO APE

Chris Borthwick

ABSTRACT

Charles Darwin's correspondence with Dr. John Down indicates that both men thought of people with mental retardation as being akin to humanity's evolutionary predecessors. Recent attempts to teach people with mental retardation to use the computer languages taught to apes show that this conceptualization of mental retardation has remained an unspoken and perhaps unformulated element in public and professional attitudes to people with disabilities ever since. Refuting this conceptualization points the way to a more productive approach to the conditions now classified as 'mental retardation.'

EVOLUTION AND IDIOTS' EARS: ANCESTRAL SURVIVALS

In Darwin's *The Descent of Man*, first published in 1871, he puts forward the proposition that some parts of human anatomy represent the attenuated relics of similar organs that had different or more important functions in the animals from whom we are descended. These relics included the ability to wiggle one's ears.

> The extrinsic muscles which serve to move the external ear are in a rudimentary condition in man... I have seen one man who could draw the whole ear forwards; other men can draw it upwards; another who could draw it backwards...

Darwin traces the form of the human ear back to the apes, and in particular studies one evolutionary peculiarity in the external ear:

> a little blunt point, projecting from the inwardly folded margin, or helix. ... The helix obviously consists of the extreme margin of the ear folded inwards; and this folding appears to be in some manner connected with the whole external ear being permanently pressed backwards... (Darwin [1876] 1989, p. 19).

Darwin traced this feature to an ancestral pointed ear, now folded over. It is now known as Darwin's tubercle. Darwin had, in his usual fashion, taken pains to collect as many accounts as possible of variant forms of the feature. In the second edition of the *The Descent of Man* in 1876 he mentioned one of his sources.

> I have myself seen, through the kindness of Dr. L. Down, the ear of a microcephalous idiot, on which there is a projection on the outside of the helix, and not on the inward folded edge, so that this point can have no relation to a former apex of the ear. Nevertheless, in some cases, my original view, that the points are vestiges of the tips of formerly erect and pointed ears, still seems to me probable (Darwin [1876] 1989, p. 20).

Two letters from kind Dr. Down to Darwin are still extant in the Darwin papers[1] and show Down also to be a precise student of ears.

> So far as I can judge from the external configuration and general feel of the ear I wrote to you about, it ... is not by any means a depraved one. Contrary to what one usually finds in the ears of Idiots there is a free lobule and the Helix and other parts are fairly developed (Down to Darwin, MS, 20 Dec. 1873).

IDIOTS, RACES AND THE EVOLUTIONARY LADDER

Having caught the attention of England's greatest evolutionist, Dr. Down naturally takes the opportunity to discuss his own theories on related subjects.

> I have been for a long time working at an antithetical subject to that which has engaged your attention; to involution rather than evolution and with results confirmatory of your teaching—
>
> In 1866, I published some observations I had made with reference to the change of race type associated with degeneration. My paper did not excite much interest at the time, but on the late visit of the British Medical Association to my establishment for training Idiots at Normansfield, Hampton Wick the interest in the question was revived (Down to Darwin MS, December 20, 1873).

One hundred and thirty years later the interest remains, because Down's paper was his classic 'Observations on an Ethnic Classification of Idiots' containing the first description of the syndrome that now bears his name. Down told Darwin:

> I showed the Members [of the BMA] several specimens of typical Mongols, the progeny of Caucasian parents. I have had negroids & Malays from like parentage (Down to Darwin MS, December 20, 1873).

Down's terms, including the 'Mongolian idiot,' were explanatory as well as descriptive. They worked from the theory of recapitulation, which postulated that:

1. Higher animals in their embryonic development passed through a series of stages representing the adult forms of their lower ancestors, and
2. Higher human races had passed through and developed beyond the stages now represented by the existing civilizations of the lower races, and
3. The evolution of animals and human races were two comparable ladders, with lower forms stopping their climb at lower levels, and
4. within a race, some individuals might slip back down to ancestral levels—'throwbacks,' or 'atavisms.'

Atavisms, in the terminology of the time, were:

> the spontaneous reappearance in adults of ancestral features that had disappeared in advanced lineages... Arrests of development represent the anomalous translation into adulthood of features that arise normally in fetal life but should be modified or replaced by something more advanced or complicated. Under the theory of recapitulation, these normal traits of fetal life are the adult stages of more primitive forms (Gould 1980, p. 136).

Down suggested that if there was a ladder of evolution, then throwbacks from superior (Anglo-Saxon) races would end up further down that ladder—on the rungs occupied by lower (colored) races. He saw his concepts as both drawing on and illuminating general evolutionary theory and the descent (or ascent) of man. In particular, he saw them as illuminating the relationship of the different races. As English Anglo-Saxons can revert to Chinese or Negros, 'These examples of the result of degeneracy among mankind appear to me to furnish some arguments in favour of the unity of the human species.' The Chinese are in Down's eyes more primitive examples of humanity, but they are on the same line of development as the English gentleman, not different and separate lineages. Different races are, in the technical language of the time, varieties, not species. As he writes to Darwin:

> Of course the argument is that if these changes of race type can be produced by degeneration there is an end to the so called species and the races are true varieties (Down to Darwin MS, December 20, 1873).

It was this liberal conclusion, not the naming of a particular variety of idiot,[2] that he saw as his achievement and it was that aspect, rather than the naming of a particular feature of the ear, that he expected to interest Darwin.

> Probably, however, all this has occurred to you before and I am only taking up your time with a thrice told tale.

From the point of view of the study of mental retardation, however, the significant point is not what conclusions Down drew about race, but the manner in which he looked to the anatomy of idiots to provide guidance on aspects of evolution. Down saw idiots as throwbacks, and thus in some ways akin to the missing link between ourselves and our remote ancestors. Darwin appears to have shared this view, which would explain why he was looking to Dr. Down's establishment at Hampton Wick to provide evidence not available so readily elsewhere; he expected the ears of idiots to show vestiges of ancestral function that the ears of normal people had evolved beyond.

In a later letter Down raised other possible reversions, describing the dissection of a microcephalic idiot of about four years of age.

> The interest about the case is mainly the condition of his ears whch [sic] had the rudimentary lobe well developed as figured in your last work. The right pinna

Idiot into Ape

was especially well marked and that I have preserved. ... On one side the Occipital lobe is completely cut off from the rest of the cerebrum by a deep sulcus such as one meets with in Anthropoid Apes...I have preserved the encephalon & medulla spinalis, right pinna and sacrum with coccyx... (Down to Darwin, MS, December 28, 1873: DAR 87, pp. 63-64).

The coccyx, or tailbone, is vestigial in both apes and men; was Down expecting it, too, to show signs of primitive descent—perhaps a tail? Such an expectation would fit with Darwin's expressed views. Elsewhere in The Descent of Man he appears to regard idiots as being in many ways linked to earlier evolutionary stages ('brute-like').

Microcephalous idiots ... are often filthy in their habits, and have no sense of decency: and several cases have been published of their being remarkably hairy. The simple brain of a microcephalous idiot, in so far as it resembles an ape, may be said to offer an example of reversion (Darwin [1876] 1989, p. 36).

Darwin's and Down's opinions on this point found ready acceptance: a decade later a popular book of lectures on mental diseases described the case of E.D., a case of genetenous idiocy:

You see [in the original lecture the patient was brought into the theatre for display] her body is squat and ugly ... She has from childhood beaten her head with her hands, as you see her now doing, just as the gorillas beat their breasts in the African woods. Her face is utterly unhuman, hence such cases have been called theroid or beast-like. The evolutionists would find many proofs of reversion to conditions common in the lower animals in her (Clouston 1887).

Darwin's ambiguous relationship at this time in his life to the concepts of development, recapitulation, and reversion has been explored by Oppenheimer (1967). Darwin wrote in *The Origin of Species*:

As the embryonic state of each species and group of species partially shows us the structure of their less modified ancient progenitors, we can clearly see why ancient and extinct forms of life should resemble the embryos of their descendants—our existing species.

As Oppenheimer points out, Darwin:

...explained embryonic resemblances on the basis of community of descent, with great profit to his own argument; but he wished to believe what von Baer had so vehemently denied, namely, that embryos could mirror the history of the race by being similar to adult, though extinct, forms (Oppenheimer 1967, pp. 251-252).

Darwin repeated and strengthened these recapitulationist ideas in the final edition of The Origin of Species in 1871. His leaning towards recapitulation also involved an openness to reversion, as in arrests of development, and thus to a belief in the primitivism of idiots.

We would today regard this linkage between individual embryo-to-adult growth and evolutionary charge as un-Darwinian, specifically because it is based on developmental imperatives rather than natural selection—because, in Bowler's words, "...it encouraged the belief that evolution shares the progressive and teleological character of individual growth" (Bowler 1988). Darwin himself, however, was not consistently Darwinian, and at the time he wrote *The Descent of Man* was flirting with Lamarckism and the inheritance of acquired characteristics (suggesting, for example, that mutilations to guinea pigs might be reproduced in their descendants [Darwin 1989, p. 63]).

Phineas T. Barnum, here as elsewhere, encapsulated an intellectual climate in a memorable fraud. The year after the publication of *The Origin of Species*, a poster for his museum showed a humanoid form with the text "WHAT IS IT? Is it a lower order of MAN? Or is it a higher order of MONKEY? None can tell! Perhaps it is a combination of both! It was captured in a savage state in Central Africa. It has the skull, limbs and general anatomy of an ORANG OUTING and the countenance of a HUMAN BEING." (Thompson 1997. p.69) It was, in fact, William Henry Johnson, a black man with microcephaly. The terms ape, primitive man, negro and idiot were collapsed into one person.[3]

IDIOTS, APES AND ALMIGHTY GOD: PRE-DARWINIAN ANALOGIES

The perception of a resemblance between idiots and primates was not dependent on the acceptance of the theory of natural selection. Dr. Samuel Howe, writing in 1848, 11 years before *The Origin of Species*, had expressed the same thought in the language of phrenology.

> Those tribes which still linger behind in savegedom show us the race not yet emerged from its youth, by reason of the great comparative activity of the animal propensities. Now it seems as if the dwarfed brain of the idiot shows us a still earlier and lower condition: it exhibits the animal man still more closely, and shows him to resemble the monkey most closely in his looks. It is not merely the *up-looking* and twinkling eye, the flattened forehead, the projecting jaws, and the

other anatomical peculiarities that give him this likeness, but sometimes, moreover, the likeness is seen in habits and actions.

...[if] this likeness is strongest in those who have very small brains, then we may suspect ... that there has been a progressive development of that organ in the progress of the race. Some of these habits seem to show the reappearance of instincts which could only have belonged to man in a low animal condition, and which have entirely died out in the race long ago ... (Howe 1858, p. 61).

Howe was a devout Christian, and believed that it would be tantamount to blasphemy to suggest that idiocy was a natural occurrence:

It seemed impious to attribute to the Creator any such glaring imperfection in his handy-work. It appeared to us certain that the existence of so many idiots in each generation *must* be the consequence of some violation of the *natural laws*—that where there was so much suffering there must have been sin (Howe 1858, p. vi).

He thus attributes every case of idiocy to such sins as intemperance, self-abuse, intermarriage of relatives and attempts to procure an abortion.

The interesting thing about the idea of the idiot as throwback is that it was common ground for two opposed conceptions of evolution and two different traditions of care for people with intellectual disability. Even more interesting, the idea in its essentials has been transmitted unaltered through every change in intellectual fashion from Howe's time to today.

It is perhaps necessary to clarify and qualify the nature of this statement. Two of the most striking things about the field of intellectual disability are that no agreement obtains about what 'intellectual disability' is, and that this is not thought to be a problem. The most recent American Association on Mental Retardation Manual on Definition states that "Mental retardation is not something you have, like blue eyes and a bad heart. Nor is it something you are, like being short or thin. It is not a medical disorder.... Nor is it a mental disorder ... [it] is not synonymous with [its] etiology ... [it] refers to a particular state of functioning ... [and] describes the 'fit' between the capacities of the individual and the structure and expectations of the ... environment" (AAMR 1992) .

While a considerable advance on previous definitions, this is both somewhat disingenuous—not all cases of mismatch between capacity and environment are intended to be comprehended by the definition— and plainly incomplete. If people are not provided with an official conceptualization in terms of what people with mental retardation have and are, this does not mean that they do not hold such concepts, only that the concepts are unarticulated and unregulated.

Terms such as 'mental retardation,' and the field that employs them, are enabled to function because the people who use them have a predefinitional concept of what mental retardation is and what it is like.[4] The retardation in question is seen as a backwards relationship along a fuzzily conceived ladder of progress.

I do not claim that the throwback model of retardation is at the forefront of people's minds, or even that many people working in the field of intellectual impairment would nominate it if prompted. Instrumentalist definitions such as the IQ measure/social skill criterion appear to have removed the need to think of any model at all in any precise formulation. The significance of the throwback model is that it can function perfectly well as an unrecognized and unformulated foundational presupposition, surfacing to written expression only when some novel setting such as experiments with ape language[5] comes along. As Darwin lamented in relation to his own theory, intermediate forms are often missing and we must attempt to plot a story from widely separated data points. Comparisons with apes occur seldom in the literature, but *any* speculations on the nature of retardation are surprisingly rare; when such speculations do infrequently occur, the ape analogy is one of the comparisons that has been most often used.

The notion of the mentally retarded person as in some vague sense an evolutionary throwback represents a basic, though often unexpressed, component of public and professional conceptualizations of the condition.

THE LADDER OF CREATION

Because the concept of relative placement is predefinitional and inchoate it is impossible to be exact about the nature of the comparisons that are being made. Down's ladder combined the progress of civilizations, races, and species and the development of individuals into a single developmental scale (Borthwick 1996). Some theories of racial development and disability have been more narrowly focused than others. Dr. Livingstone, the African explorer, was, among his other duties, looking for a tribe that stammered, to test out the hypothesis that stammering was a disease of higher civilizations (Rockey 1980, p. 26), and other explorers looked for a tribe that used sign language, to test out the hypothesis that deaf sign was a primitive survival (Peet [1848] 1997). There is, however a general consistency of approach. The relative importance attached to each element of the scale has varied from time to time across the last 150 years, and the use of any element has often tended to bleed across into the others.

The guiding metaphor remains that of the ladder of creation, and this metaphor has been enormously and continuously influential. Such fathers of modern psychology as Yerkes structured their professional lives, and modern psychology, around it.[6]

> Yerkes believed that simpler forms of life might be viewed as living fossils and began the task of tracing the phylogenetic development of human intellectual capacity back through lower forms of life to its origins with studies of the behavior and 'mental life' of frogs, jellyfish, crustaceans, worms, mice, crows, pigs, raccoons—and eventually insane persons, schoolchildren, soldiers, orangutans, chimpanzees and gorillas (Reed 1987).

Yerkes' work on the orangutan and the insane might be compared with other contemporary efforts to link developmental lineages. There seemed to be an agreement on the significance of the orangutan. In 1924 Dr. James Kerr mentioned that "the mongol takes after the urang and sits cross-legged" (Booth 1985) while in his *The Mongol in Our Midst* Crookshank, also writing in 1924, placed the ape on the next level down.

> The cases in which the mental defect is least pronounced tend, morphologically, to resemble children of the Mongolian race while marked cases of this type appear to be simian, and among the anthropoid apes, [resemble] the orang (Brousseau 1928, p. 13).

One step down the ladder took you to the Mongolian, two steps to the orangutan, or the chimpanzee.[7]

Reversionary views did not find universal favor among the field. Sherlock, writing in 1911, had already criticized Darwin's approach in this area:

> In 1867 ... biologists were engaged in a lively discussion of the evolutionary theory which Darwin had promulgated a few years before, and an intemperate zeal led the adherents of the new school of thought to conclude, somewhat rashly, that microcephalic idiots represented a return to the simian stage in the geneology of human beings—a view that Darwin himself seems to have accepted.... It soon became apparent that this explanation was not particularly well founded ... (Sherlock 1911, p. 217).

Sherlock did not, however, regard the issue as entirely settled: he also said that "to what extent (developmental errors) are due to a process of reversion remains open to question." and went on to report and reference an 1895 dissection of two microcephalics, 'Fred' and 'Joe', where 'Fred' was 'a case of partial atavism' and the brain of Joe 'displayed simian features.'

In the fifth edition of his work *Mental Deficiency* Tredgold (1929) rejected the notion of atavism (evolutionary regression) in favor of a more narrowly conceived 'arrest of development' and his views have largely prevailed. It should be noted, however, that Crookshank's introduction of the term 'simian crease' to describe a common feature of Down syndrome anatomy is still in use today, and that Tredgold's rejection of atavism was in part eased by his continuing belief that fetal life progressed through earlier (racial) types of mankind and so an arrest of development could mean that the mentally deficient person did indeed carry Mongolian characteristics just as if they were an atavism.

As the century progressed, however, specialization increased, and the wide-ranging speculation of such amateur evolutionists as Crookshank (and Dr. Down) began to go out of fashion. Cyril Burt, another of the pioneers of psychometrics, noted the popularity of the 'throwback' belief in his 1937 book *The Backward Child* only to dismiss it as unscientific. He recorded:

> the round receding forehead, the protruding muzzle, the short and upturned nose, the thickened lips, which combine to give the slum child's profile a negroid or almost simian outline. Once again, these are the very defects which are singled out as the stigmata of degeneracy and supposed to indicate a throw-back to some primitive stock: and often the primitive, impulsive conduct of these youngsters seems to bear out this notion. "Apes that are hardly anthropoid" was the comment of one headmaster ...
>
> In point of fact, however, there is no need to invoke the doctrine of ancestral reversion ... Both the appearance and the behavior may be explained as acquired rather than inherited characteristics (Burt 1937, p. 186).

Burt, while a strong and vocal believer in national degeneration, may have felt that the random genetic lottery implied by the idea of 'throwbacks' was inconsistent with his belief in eugenics, which relied for its effect on intelligent parents tending to breed intelligent children and stupid parents stupid children. His belief in the ladder of creation remained.

This view of evolution as progressive—as developmental—retains a strong grip on many important areas of modern psychology. Jensen, for example, accepts the 'phylogenetic hierarchy' as a foundation of his belief in the supremacy of the *g* factor in intelligence testing (Jensen 1978). Gould (1996) and Borthwick (1979) have commented on the particularly Nineteenth Century quality of this aspect of Jensen's work. It is not clear whether a belief in the validity of psychological test rankings leads to a belief in the ladder of creation, or vice versa; it is clear that the two are not simply consistent but strongly mutually reinforcing.

The ladder image says both that some people are higher or lower than others and that they are capable of moving up or down only along one narrow pathway. Applied to intelligence testing, it necessitates the belief that in both the individual and the species a single unilinear scale represents the only significant axis of development. It thus necessitates the further assumption that people with mental retardation are like apes: if there is only one possible scale of difference, then things that are different from the same thing must resemble each other.

Similarly, the tradition of children's education and the special education that arose from it was created by men who believed in recapitulation. Pestalozzi and Froebel, and after them Piaget and Freud, believed that there was a developmental ranking that paralleled evolution; as Pestalozzi said, "The child masters the principle of cultivated speech in exactly the same slow order as Nature has followed with the race" (Gould 1977, p. 48).

THE APE ANALOGY IN THE POSTWAR PERIOD: FOLLOWING THE THREAD

By the 1960s, disciplinary boundaries had tightened further and the evolutionary speculations of Dr. Thomas Merton as to why white mongoloids were more like orangutans while black mongoloids were more like gorillas (Merton 1968) were the privately-printed hobbyhorse of a country medical practitioner.

As late as 1967, however, a symposium of medical specialists gathered to mark the centenary of Down's original paper could devote a significant part of their discussion to reviving studies (from 1928) of the existence in persons with Down Syndrome of "tuberosities of the medial part of the peduncle of the cerebellar flocculus." These are said to disappear before birth in 'normal' infants but are present in mature orangutans and chimpanzees (Wolstenholme and Porter 1967). These features were said to support the position that Down Syndrome was an 'arrest of development' involving a return to phylogenetic roots.

That seminar was itself something of a throwback; such insights into professional conceptualizations are increasingly rare, and speaking about such comparisons has certainly become less common in recent decades, in part because the equivalence of racial and evolutionary progress has become more contestable. While atavistic models are articulated with comparative infrequency, however, it is perhaps more important that in many areas we continue to act on the same principles.

It might be thought that as the throwback school of evolutionary theorizing went increasingly out of fashion after World War II, the view of mentally retarded people as somehow ape-like would simply fade away. This has not happened: as long as the assumption of a unilinear intelligence remains, it cannot happen. The association remains, but has gone underground. It is not drawn upon overtly, but is nonetheless enormously influential. The power of the concept is most clearly seen in the manner in which it can operate freely in the presence of overt denial. Researchers can specifically deny making comparisons between mentally retarded people and apes in the same publications in which they specifically compare mentally retarded people and apes. Deich and Hodges, for example, say that some people "may erroneously assume that we are drawing a parallel between the chimpanzee and the non-verbal retarded child; however, no comparison is intended." This is plainly untrue: exactly that comparison is intrinsic to their experiments comparing the communication of people diagnosed as having intellectual impairment with the communication of apes. Similarly, Romski has vehemently denied any intention of making such comparisons (personal communication 1998).

The continued popularity of this line of experiments provides the clearest indication of the continued influence of the notion of retardation as reversion to a lower level of the evolutionary scale.

> It had been noted that the performance of the people with mental retardation on a limited range of tests did in fact show some similarities to those of apes.
>
> In 1935 Gottschaldt studied a group of one hundred hospitalized children between the ages of two and ten. He divided them into four groups in descending order of intelligence: normal, feeble-minded, imbecilic, and idiotic, and set them the same, or nearly the same, problems Kohler had set his chimpanzees. The

results were comparable, and so was the order of difficulty which emerged ... (Viaud 1960, pp. 44-45).

Apes are like two-year-old children, the argument went, because they can use a restricted vocabulary for a restricted range of purposes. Language experiments involving attempts to extract expressive language therefore used the conceptualization of apes as like children. However, apes are not like children, because children can speak and apes cannot. Apes are therefore more like the other group that is 'like' two-year-old children—people diagnosed as mentally retarded who have limited speech.

APE LANGUAGE AND MENTAL RETARDATION: HISTORY

Yerkish, a 'computer-controlled language system for investigating the language skills of young apes,' was devised in the Yerkes Primate Center in the 1960s. The late 1960s and early 1970s saw the introduction of these 'laboratory' languages to hearing non-speaking children in the United States (Savage-Rumbaugh et al 1973). These programs were directly based on the use of such communication by chimpanzees. The title of one book on the topic, *Language Intervention from Ape to Child* (Rabusch et al. 1984), reflected the attempts to use these strategies with non-speaking 'retarded' human infants.

> Savage-Rumbaugh has her detractors, but she also has her supporters, some of whom have applied her techniques to communications research with humans. For instance, the keyboard technology has provided profoundly mentally retarded kids with a means of communicating. J.J., 18, can't speak, but after a decade of working at the language center with a keyboard like Kanzi's, he can finally communicate. ... [Romski said that] "I think that what's so powerful about it for the children, is that when, for example, JJ is able to tell you something through the use of his board and, umm, speak to you that way, that you see something about what he knows and otherwise, you don't have that window. You just think he's a child that has mental retardation and sort of grunts and vocalizes and smiles at you, maybe laughs with you, but you don't know what's really in his mind. But when he uses specific symbols in specific ways, you have a sense of what he knows about his world. It gives you a window into their minds that otherwise you wouldn't have" (Johnson 1995).

The use of these devices having been established in people with retardation, the devices are then used by people working in primate research, further strengthening the association.

Panbanisha, a Bonobo chimpanzee who has become something of a star among animal language researchers, was strolling through the Georgia woods with a group of her fellow primates—scientists at the Language Research Center at Georgia State University in Atlanta. Suddenly, the chimp pulled one of them aside. Grabbing a special keyboard of the kind used to teach severely retarded children to communicate, she repeatedly pressed three symbols—"Fight," "Mad" and "Austin"—in various combinations (Johnson 1995).

Romski and her colleagues claimed that the establishment of a conditional association between food and its symbol had formed the foundation for symbolic communication skills in two chimpanzees, and went on to describe "the successful adaptation of this language-relevant research effort to language intervention in five case studies of adolescents with severe retardation." (Romski, Sevcik and Pate 1988, p. 95) The term 'successful' may perhaps be queried; a further study reported in the same article involved teaching 20 food-related words to three of four persons diagnosed as severely retarded through the use of 57,300 trials in over 1,400 half-hour sessions. There was no control group, and no alternative interventions were trialed.

A 1977 study by LaVigna recorded the teaching of written words to three mute autistic adolescents using a procedure based on "Premack's language training with chimps." (La Vigna 1977) This was regarded as evidence that "written words may provide a viable communication system for the mute autistic". While the observation is true and valuable, LaVigna's experiment cannot be said to have illuminated it. In this study, again, there was no control group and no alternative interventions were trialed. As the three subjects required some 4,400 trials to learn three words, the adjective 'viable' seems excessive.

Another study by Carrier and colleagues (1978) was strikingly more efficient, teaching basic reading skills (with a limited vocabulary) to 14 children diagnosed as severely retarded. The Carrier experiment shows the essential problems with ape-based treatments. Premack taught apes to use symbols for communication; Romski drew on that to teach people diagnosed as retarded to use symbols for communication; LaVigna and Carrier substituted words for symbols; gradually, replacing first the axe haft and then the axe head, the methods approach those that would have been thrown up more quickly had researchers turned first to the experiences of children in normal schools. It is easier and faster to reach backwards from our own capacities than forward from those of Bonobos.

Whatever the outcomes for the trial subjects themselves, and whatever the reliability of the findings regarding the language capacities of apes, the conclusions drawn from the specifically retardation-focused aspects of these studies have had several positive aspects. LaVigna observed that

> It is conceivable that the written word involves less perceptual and behavioral complexity than the spoken word, and therefore may present fewer problems in training (LaVigna 1977, p. 147).

This concept provided a valuable stimulus to the use of non-vocal systems. The prominence of the field gave impetus to ventures in the 1970s and 1980s to provide communication equipment to people with intellectual disabilities. Many users have benefited from this approach.

While the use of strategies based on ape language has had its positive aspects, it has also had drawbacks. The climate of suspicion around ape language has as one consequence that people diagnosed as retarded who are learning to communicate are sometimes regarded as falling on the animal side of Morgan's famous rubric on anthropomorphism—"In no case can we interpret an action as the outcome of a higher psychical faculty if it can be interpreted as one which stands lower on the psychological scale" (Morgan 1894, p. 53). This means that disabled people sometimes have to meet the same standards of evidence to demonstrate their individual capacity for language as would be called for in chimpanzee experiments. It is not always recognized that Morgan's rule is not an appropriate tool when dealing with human beings.

APE LANGUAGE AND MENTAL RETARDATION: CONCEPTUAL FOUNDATIONS

Even though the outcomes for people diagnosed as retarded, as a class, may have been at least in part positive, it is necessary to ask why comparative experiments of this nature were carried out, why the analogies were thought to be plausible, and what their theoretical foundation was. Any search for a theoretical base has to explain not why humans were thought to be comparable to apes, but why analogies drawn from primates were thought to be more relevant to people diagnosed as retarded than they were to ordinary people.

At first sight the arguments for drawing on ape studies seem to be so reliant on false assumptions and non sequiturs as to raise questions about whether they were sincerely held. Romski, for example, suggested that:

> ... typical children learn to speak and youngsters with severe spoken language impairments frequently do not. Whatever language they learn must be taught. Non-human primate research can provide a model ... (Romski 1989).

The transition from speech to 'spoken language impairment' to 'language' covers an unargued assumption that people without speech have no language—in the same way, the next line implies, that apes without speech have no language. It is exactly that similarity which is at issue and requires justification. Only a very fundamental and unquestionable presupposition of apeishness could bridge that logical gap, and the weakness of the argument is a demonstration of the strength of the prejudice.

If researchers looked to apes for these analogies rather than to ordinary people, did this imply that people diagnosed as retarded were thought to be closer to apes than they were to ordinary people? Some writers suggest that both groups held positions on a single scale of mental development (in this case, 'mental age'). Deich and Hodges, the authors of one of the first books on non-speech communication therapy with mentally retarded people, said that:

> ... we can take the findings from subhuman language learning and apply them to humans who are impaired or delayed in language. ... For example, if chimps can learn to communicate non-vocally with humans, we may predict that retardates of roughly comparable mental age can also learn to communicate using the same technique (Deich and Hodges 1977).

The comparison between apes and people diagnosed as retarded is supported, of course, by resemblances between the performances of each group. Apes cannot speak; people with severe retardation often cannot speak.

> If non-vocal response modes provide an alternative for chimpanzees who do not have the physical ability to produce speech, it seems possible that such an approach might have applicability for communication development in low-level retarded [humans] (Kunz, Carrier, and Hollis 1978).

Such a likeness does not, however, necessarily lead one to comparisons with apes, rather than other groups who lack the physical ability to produce speech—cows, say, or rocks—unless a further assumption is made: apes and people with severe retardation have similar minds.[8] This assumption may remain, as in the article by Kunz and colleagues, unstated, but it is no less influential for that. Indeed, the fact that writers such as Kunz assume correctly that the reader will fill in the necessary assumption illustrates how widespread that assumption is in our society. It is an assumption that draws on a commonly held framework of evolutionary progression, on the notion of a ladder.

The ladder can extend even further, below the apes: In a 1949 experiment Fuller sought to have an 18-year-old 'vegetative human organism'[9] raise his right hand by denying him food except when reinforcing suitable movements.

> While of normal human parentage, this organism was, behaviorally speaking, considerably lower in the scale than the majority of infra-human organisms used in conditioning experiments ... dogs, rats, cats.... Perhaps by beginning at the bottom of the human scale, the transfer from rat to man can be effected (Fuller 1949, p. 588).

We are back with Barnum. Lower humans might overlap with the highest animals, and the very lowest humans might overlap with the lower animals and might be able to learn similar skills. It is this vision that underlies the connection between ape language and language training for people with disabilities.

> "It required just a few second's thought to realize that if we could learn how language develops in a chimpanzee, we would surely have a leg up in learning how to cultivate language in mentally retarded children," Duane now says, modestly (Savage-Rumbaugh and Lewin 1994, p. 181).

The modesty may be appropriate, as the thought is a total non- sequitur, one that fits so well into public preconceptions that it went uncorrected and unnoticed by academics, research funding agencies, and communication therapists alike.

But why have apes and people diagnosed as retarded been thought to have similar mentalities? Gottschaldt's results are unsurprising, given that chimpanzees and children with motor impairments are equally disadvantaged on tests which require fine motor skills and a pincer grip. More generally, such comparisons are able to document resemblances only because they deal with a strictly limited scale of achievements, or more properly deficits, that have been as far as possible separated from the physical capacities of the participants. A BMW with defective spark plugs may perform like a farm cart; its speed may even improve if it is treated like a cart and yoked to an ox. It would nonetheless be profoundly misleading to place the two vehicles in the same category, or to draw conclusions about either maximum speeds or appropriate repairs from one's experience of carts. The possible efficacy of communication devices with people like J.J. could, similarly, have been more easily and more illuminatingly deduced from their usefulness with other people. If we ourselves can use typewriters, under what circumstances can J.J. use something similar?

The governing ideas of the mechanism of evolution have changed radically and frequently. Haekel's recapitulationism was followed by Bolk's neoteny, and progressive development was eventually ousted by the incorporation of Mendel's ideas into the modern Darwinian synthesis. The fancied resemblance of people with intellectual impairment to apes has remained a constant across all periods and all theories.

THE SURVIVAL OF THE DEVELOPMENTAL ANALOGY

It is clear to us now that the racial analogies drawn by Down between lower races and people diagnosed as mentally retarded were incorrect and irrelevant, and we may congratulate ourselves for our insight. We may, however, like Butler's Hudibras:

> Compound for sins we are inclined to
> By damning those we have no mind to.

While racial comparisons are out of favor, the analogy between apes and people diagnosed as mentally retarded has, as we have seen, surprising vitality. It is, nonetheless, fatuous. We now know more about brain structure and function than we did in Darwin's time, and what we know is that the programming of all ape brains is very different from all human brains. While atavism in some physical features does occasionally occur in humans, this does not provide an explanation for the phenomenon of mental retardation. There is every reason to believe, for example, that our ancestors possessed only 46 chromosomes.

There are strong arguments for believing that the study of the communication of apes may cast light on human communication.[10] That is not the issue. There is no reason whatsoever to suppose that people diagnosed as retarded have minds that resemble those of other species any more than do those of other people. There is, after all, no organic similarity at all. Darwin was wrong; it is impossible to find in these people physical analogies to ancestral species. Nobody now suggests that such things as hairiness or ear shape can be used to identify people with disabilities: The physical aspect of the throwback myth has simply ceased to exist, without, apparently, affecting the conclusions that were drawn from it.

In the *The Descent of Man*, Darwin wrote:

> In the next chapter I shall make some few remarks on the probable steps and means by which the several mental and moral faculties of man have been gradually evolved. That such evolution is at least possible, ought not to be denied, for we daily see those facilities developing in every infant: and we may trace a perfect gradation from the mind of an utter idiot, lower than that of an animal low in the scale, to the mind of Newton (Darwin [1876] 1989, p. 54).

If you believe, however subliminally, in a ladder of creation—a two-dimensional gradient from higher to lower—then you must believe that people diagnosed as mentally retarded are closer to apes, and to children, and to rats, than normal people are. There is simply nowhere else for them to stand. Most people, most psychologists and most people working in the field of intellectual disability do at some level of their thought believe just that.

Evolutionists today tend to stress the other, more 'Darwinian' Darwin, who wrote in his notebooks that it was absurd to talk of one animal being higher or lower than another. (Gould 1989, p. 256). Darwin was a very great man with an enormous capacity to withstand the consensus of his environment, and the fact that he was unable to retain this insight in his later writings is a testimony to the strength of the continuous pressure of the countervailing illusion in our culture.

Darwin was right the first time. There is no ladder of creation. People diagnosed as mentally retarded are not like animals, or our ancestors, or anything other than ourselves, differenced. The minds of people diagnosed as retarded are not smaller, or lesser, or less complicated than those of other people, any more than their brains are or their DNA is. Their minds may be more complicated, in the same way that a broken plate is more complicated than an intact plate. It is precisely the influence of the evolutionary and developmental model that prevents researchers and readers from seeing this obvious point. Unless one does believe that people diagnosed as mentally retarded are 'throwbacks' or 'missing links'—that there is a ladder of creation on which they stand lower—then there is simply no defensible reason to compare their communication with that of apes.

The notion of the mentally retarded person as in some vague sense an evolutionary throwback represents a basic, if often unexpressed, component of public and professional conceptualizations of the condition. It would be greatly to the benefit of people with disabilities if this underlying belief in their kinship with apes could be uncovered, exposed and refuted. Even such a reform would leave us (and Darwin) with the other prop of the developmental approach—the notion of the mentally retarded person as a child, having a mental age of two or four or six-still in place. Again, this comparison is founded on no concrete evidence. Neurology has uncompromisingly established that the brain of a child is very different from the brain of an adult. It is still growing in size until the age of one, still establishing new neuronal connections till the age of two, reaching peak metabolic rates at the age of four, and still establishing fundamental elements of its eventual structure until adolescence (Upledger 1997).

As with apes, so with children; if different structures produce similar outcomes, they must do so by different means that have no necessary application to each other's problems. The concept of mental age has been a protracted and unhappy detour from the investigation of the nature of brain function damage in people diagnosed as retarded.

All these attempts to reason about deviant mental functioning by analogy with other low-scoring groups suffer from the same conceptual defect. Things that are different from the same thing do not necessarily resemble each other. People diagnosed as having mental retardation do not necessarily resemble each other. It is only the assumption of a unilinear scale that legitimizes the belief that there is a single state that is mental retardation. If there are instead a multiplicity of unrelated ways in which minds can differ from each other, then a person's mental functioning need not be constrained to a rung in a ladder but can lie at any point in n-dimensional mindspace, unusual mental functioning can lie in any direction from the clustering that represents the statistically normal, and there is no a priori reason to expect that any particular example of unusual mental functioning would lie close to any other.

If we can learn anything from experiments with apes that is relevant to people diagnosed as mentally retarded, it might be the lesson that emerged from Savage-Rumbaugh's work involving macaques learning to play computer games.

"If you view monkeys as operating on basis stimulus and response—and a lot of people still do that—then that's all you're going to get out of them." Rumbaugh says. "If you don't even look for intelligence, then there's no way you're going to see it" (Blum 1994, p. 38).

The problem is that the real educational lessons to be learned from apes are still not considered relevant to people with disabilities.

Could it be that our assumptions of limited abilities has inextricably led to circumscribed learning, even though we had been attempting to press the skills of [the apes] to the limit? ... If I had simply assumed that Kanzi was able to learn as humans do, might not those assumptions have produced a very different kind of ape? (Savage-Rumbaugh and Lewin 1994, p. 138).

If we can assume that apes can learn as humans do, why is it so difficult to believe that other humans can?

ACKNOWLEDGMENTS

I wish to express my gratitude for the kind assistance given to me by Adam Perkins and Alison Pearn of the Darwin Correspondence Project, Cambridge University Library Archives, which has copyright in the unpublished letters by Charles Darwin cited above.

NOTES

1. Darwin's replies, unfortunately, do not appear to have survived.
2. The term *idiot* is of course now properly held to be offensive, as are many of the other terms quoted elsewhere in the paper. In Darwin and Down's time these were the correct medical terms, and their use in this paper is intended to cover the population the Victorians understood as falling under these terms.
3. Barnum later blended the terms *idiot* and *primitive race* in a similar scam, advertising other microcephalics as 'Aztecs' (Kunhardt 1995), but in this he was simply following educated usage: the head shapes found on Aztec artifacts were widely seen as shockingly deviant.
4. These assumptions emerge most clearly in the criteria for the sub-diagnosis 'Unspecified Mental Retardation,' where "there is a strong presumption of Mental Retardation but the person is untestable by standard intelligence tests [because they are] ... too impaired or too uncooperative to be tested" (DSM-IV). Considered simply in terms of the classification system, this is meaningless. A presumption of 'Mental Retardation' must, in terms of the criteria, mean a presumption that (inter alia) a person has a measured IQ of below 70 on a standard intelligence test. The diagnosis of 'Unspecified Mental Retardation' can be supported only by reference to the unwritten but well

understood concept of 'people who look and behave like other people diagnosed as mentally retarded.'

5. See below.

6. Modern psychology is not, of course, a unitary enterprise, and different branches of the field hold different theoretical relationships with these ideas; a full taxonomy, however, would lie outside the scope of this paper.

7. Crookshank himself added other levels for Negroids and gorillas, but the principle is the same.

8. Or, of course, that both groups had similar physical incapacities; but that conclusion has never been accepted, or even contemplated. Of the four people diagnosed as being severely mentally retarded in the 1988 Romski and colleagues study one was deaf, one was a quadriplegic, two were in wheelchairs, and all were institutionalized, but none of these handicaps were thought to cast any doubt on the experimenters' assumption that any difficulties in complying with the experimental protocol should be put down to cognitive incapacity.

9. It is impossible to be certain without more information, but from the account given it seems likely that the teenager concerned had cerebral palsy, affecting both his speech and his ability to move his arms.

10. Although Preuss (1995) warns that exactly the assumption that there is a unilinear scale of neural organization that underlies both the 'phylogenetic hierarchy' and 'the ladder of progress' may mislead researchers into assuming more commonality between human and ape than is in fact the case.

REFERENCES

American Association on Mental Retardation (AAMR). 1992. *Mental Retardation— Definition, Classification, and Systems of Supports*, 9th ed. Washington, DC: Author.

Blum, D. 1994. *The Monkey Wars*. New York: Oxford.

Booth, T. 1985. "Labels and Their Consequences." In *Current Approaches to Down Syndrome*, edited by D. Lane and B. Stratford. London: Cassell.

Borthwick, C. 1979. "Intelligence, Learning and Evolution: a Note." Pp. 174-183 in *Melbourne Studies in Education*, edited by S. Murray-Smith. Melbourne: Melbourne University Press.

_____. 1996. "Racism, IQ and Down's Syndrome." *Disability and Society* 11(3): 403-410.

Bowler, P. 1988. *The Non-Darwinian Revolution*. Baltimore, MD: Johns Hopkins.

Brousseau, K. (rev. Brainerd, H.). 1928. *Mongolism: A Study of the Physical and Mental Characteristics of Mongol Imbeciles*. Baltimore: Williams & Wilkins.

Burt, C. [1937]. 1961. *The Backward Child*. London: University of London Press.

Clouston, T. 1887. *Clinical Lectures on Mental Diseases* (2nd ed.). London: Churchill.

Darwin, C. [1876]. 1989. *The Descent of Man, and Selection in Relation to Sex* (2nd ed., revised), edited by P. Barrett and R. Freeman. London: William Pickering.

Deich, R., and P. Hodges. 1977. *Language Without Speech*. London: Souvenir Press.

Fuller, P. 1949. "Operant Conditioning of a Human Vegetative Organism." *American Journal of Psychology* 62: 587-590.

Gelb, S. 1995. "The Beast in Man: Degenerationism and Mental Retardation, 1900-1920." *Mental Retardation* 3(1): 1-9.
Gould, S.J. 1977. *Ontogeny and Phylogeny*. Cambridge, MA: Harvard University Press.
———. 1980. *Ever Since Darwin*. London: Penguin.
———. 1989. *Wonderful Life*. New York: Norton.
———. 1996. *The Mismeasure of Man* (rev. ed.). London: Penguin Books.
Howe, S. 1858. *On the Causes of Idiocy, Being the Supplement to a Report by Dr. S. G. Howe and the Other Commissioners Appointed by the Governor of Massachusetts to Inquire into the Condition of the Idiots of the Commonwealth*. Edinburgh: Machlachlan and Stewart.
Jensen, A. 1978. "The Nature of Intelligence and its Relation to Learning." *Melbourne Studies in Education*. Melbourne: Melbourne University Press.
Johnson, G. 1995. "Chimp Talk Debate: Is it Really Language?" *New York Times*, June 6, : Section C, 1.
Kunhardt, P., P. Kunhardt, and P. Kunhardt. 1995. *P.T. Barnum: America's Greatest Showman*. New York: Knopf.
Kunz, J., J. Carrier, and J. Hollis 1978. "A Nonvocal System for Teaching Retarded Children to Read and Write." In *Quality of Life in Severely and Profoundly Mentally Retarded People: Research Foundations for Improvement*, edited by C. Meyers. Washington, DC: American Association on Mental Deficiency.
LaVigna, G. 1977. "Communication Training in Mute Autistic Adolescents Using the Written Word." *Journal of Autism Child Schizophrenia* 7(2): 135-149.
Merton, T. 1968. *Mankind in the Unmaking: The Anthropology of Mongolism*. Fairlight, Australia: Fairlight Press.
Morgan, C. L. 1894. *An Introduction to Comparative Psychology*. London: E. Arnold.
Oppenheimer, J. 1967. *Essays in the History of Embryology and Biology*. Cambridge, MA: MIT Press.
Peet, H. [1848]. 1997. "Notions of the Deaf and Dumb Before Instruction, Especially in Regard to Religious Subjects." *American Annals of the Deaf* 142(3): 8-17.
Preuss, T. 1995. "The Arguments from Animals to Humans in Cognitive Neuroscience." In *The Cognitive Neurosciences*, edited by M. Gazzaniga. Cambridge, MA: MIT Press.
Rabush, D., Lloyd, L., and Gerdes, M. 1984. *Communication Enhancement Bibliography*. East Lansing, MI: Communication Outlook.
Reed, J. 1987. "Robert M. Yerkes and the Mental Testing Movement." In *Psychological Testing and American Society, 1890-1930*, edited by M. Sokal. New Brunswick, NJ: Rutgers University Press.
Rockey, D. 1980. *Speech Disorders in Nineteenth Century Britain*. London: Croom Helm.
Romski, M. 1989. "Two Decades of Language Research with Great Apes." *ASHA* 31(5): 81-82, 38.
Romski, M., R. Sevcik, and J. Pate 1988. "Establishment of Symbolic Communication in Persons with Severe Retardation." *Journal of Speech and Hearing Disorders* 53: 94-107.
Savage-Rumbaugh, S., and R. Lewin. 1994. *Kanzi: The Ape on the Brink of the Human Mind*. New York: John Wiley.

Sherlock, E. 1911. *The Feeble-Minded*. London: Macmillan.
Thompson, R. 1997. *Extraordinary Bodies: Figuring Physical Disability in American Culture and Literature*. New York: Columbia.
Tredgold, A. 1929. *Mental Deficiency* (5th ed.). London: Bailliere, Tindall & Cox.
Upledger, J. 1997. *A Brain Is Born: Exploring the Birth and Development of the Central Nervous System*. New York: North Atlantic Books.
Viaud, G. 1960. *Intelligence: Its Evolution and Forms*. London: Hutchinson.
Wolstenholme, G., and R. Porter. 1967. *Mongolism: In Commemmoration of Dr. John Langdon Haydon Down*. London: J & A Churchill.

HOW THE NEWS FRAMES DISABILITY
PRINT MEDIA COVERAGE OF THE AMERICANS WITH DISABILITIES ACT

Beth Haller

The 1990 Americans with Disabilities Act embodied a new civil rights paradigm that put into place an empowering message about people with disabilities that sharply contrasted with the stigmatizing stereotypes and myths about disability of the past. However, as a new piece of government legislation it needed a messenger to convey its intent to the public. That messenger was the news media. Their role as societal transmitters of important disability information allowed for the investigation of what messages the news media gave the American public about disability issues in the early 1990s.

Through a content analysis of stories and photos on the ADA in the nine largest U.S. newspapers and the three major news magazines, this study sought to understand how media messages "constructed" disability in news stories and photographs. It is crucial to consider these mainstream news representations because news media make people aware of and characterize social issues, which McCombs and Shaw confirmed as the agenda-setting function of media (1972).

McCombs and Shaw believe the concept of how media frame news events is germane to agenda setting. "Both the selection of topics for the news agenda and the selection of frames for stories about those topics are powerful agenda-setting roles and awesome ethical responsibilities" (McCombs and Shaw 1992, pp. 820-821). How the attributes of a new social issue embodied in the Americans with Disabilities Act are played in news coverage could sway public opinion about both the Act and people with disabilities in general.

Because the ADA embodied the newer disability rights perspective, it was an excellent topic to investigate whether the rights perspective was slowly pushing its way into any media coverage. The passage of the ADA was analogous to Thomas Kuhn's concept of a paradigm shift in scientific discovery. The Americans with Disabilities Act represented a point in U.S. history in which the categorization of people with disabilities was shifting to the disability rights perspective. This contrasted with the way the media had traditionally covered people with disabilities and their issues. Much past media coverage had been imbedded with a medical or social welfare perspective in which disability was seen as a physical problem residing within individuals (Scotch 1988).

The disability activists advocate a perspective in which disability is understood as "socially constructed"—a phenomenon in which society has had architectural, occupational, educational, communication and attitudinal barriers to prevent people who are physically different from being totally integrated (Liachowitz 1988). In the rights perspective, physical difference is acknowledged, and even celebrated as ethnicity might be by some, but the focus is away from the disabled individual as the problem and on society's structures instead.

Higgins (1992) said we as a society "make disability" through our language, media and other public and visible ways. He argued also that in our culture "we present disability as primarily an internal condition that estranges disabled people from others." Therefore, studying news images helps us understand the media's role in "constructing" people with disabilities as different and their role in framing many types of people who may not fit with mainstream representations.

However, the story of the ADA was not about disability issues alone; it represented an intersection of government, business, and disability concerns. In the development of the ADA in the late 1980s, many members of the U.S. business community began to voice a countermessage about the negative impact of the legislation upon them. That rhetoric became part of the media story about the ADA.

For example, the U.S. Chamber of Commerce criticized provisions that dealt with the employer-employee relationship. In testimony before the Senate Committee on Labor and Human Resources in May 1989, Zachary Fasman of the U.S. Chamber of Commerce asked that all references to employment be stricken from Title I of the Act, questioning the unclear language there. He also said the definition of "reasonable accommodation" that employers must provide for people with disabilities was too broad and unnecessary. Fasman also questioned the idea of "essential function" in which someone is considered a qualified applicant if he or she can perform the essential function of a job with or without reasonable accommodation (Fasman 1989).

Transportation companies and theater owners also lobbied vigorously against parts of the ADA. In testimony before the Senate Subcommittee on the Handicapped in May 1989, a spokesperson for the Greyhound Lines Inc. said the Act would doom the company. Theater owners supported the part of the legislation that dealt with making any newly constructed buildings fully accessible to people with disabilities, but they lobbied for wheelchair accessible seating to be near an exit only. The group also wanted existing theaters to be exempt from the ADA, saying that inner-city theaters with marginal profitability could not afford to be renovated. They also argued that some jobs in theaters could not accommodate people who use wheelchairs (Green 1989).

This study argues that the media coverage of the ADA constructed several new societal representations of people with disabilities and disability issues in part because of the business and government perspectives that made it into the ADA stories. Newer images of people with disabilities, both positive and negative, entered U.S. cultural narratives:

- That people with disabilities deserve governmental civil rights protection.
- That people with disabilities cost society money.
- That people with disabilities deserve legal recourse to remedy societal discrimination.

But the findings show that these news narratives were not all presented equally. In addition, the years since the implementation of the ADA has seen a backlash against some disability accommodation because the business community's argument about the costliness of disability has taken hold in with some in American society.

REVIEW OF LITERATURE

The voice of the business community in the ADA stories obviously reflects the paradigm of capitalism in the United States, and journalists must ply their trade in this society. However, journalists were faced with a three-way pull in covering the ADA story. Mainstream journalists generally accept much government legislation unquestioningly. They also tend to support full civil rights for oppressed groups, exemplified by their support of African Americans in the 1960s. But journalists also work for businesses, and the U.S. media represent some of the largest and most profitable businesses in the United States. So most journalists readily accept capitalism.

As Gans (1980) has argued, news media embody a belief in the goodness of a free market economy. In a more critical approach, Dines (1992) has called the media "capitalism's pitchmen" because of the conservative nature of the sources they use. Her content analysis of the "voices" on network news illustrated that white, male conservatives speak most often and the perspective of the Left gets little attention.

Because the ADA was federal civil rights legislation, it forced the news media to look at people with disabilities as having minority group status and deserving full civil rights. Based on their history covering government activities, the media accepted this frame of representation because the federal government gave it to them. Linsky (1986) confirmed this interplay of government, media, and outside groups in setting new public policy agendas. The mainstream news media serve popular imagination as the "watchdogs" of government through investigative reporting. But more often than not studies show they are compliant vehicles for the rhetoric of the federal government. For example, Olien, Tichenor and Donohue showed that the media lean in favor of the status quo and the "mainstream" when covering public protests. That study found the media are watchdogs on behalf of the mainstream groups. "Media report social movements as a rule in the guise of watchdogs, while actually performing as "guard dogs" for the mainstream interests" (Olien, Tichenor and Donohue 1989, p. 24).

In the case of ADA coverage, that tendency served the interests of the disability rights movement. With little knowledge of the disability rights agenda, the news media had to rely on governmental rhetoric and disability sources to tell the ADA story. And much of the governmental rhetoric had been fashioned by activists from the disability community, unlike disability legislation of the past. (DeJong 1993). The disability rights movement broke apart the stigmatizing stereotypes of the past and moved from an invisible and marginalized status into the mainstream.

Gusfield (1981) developed a useful framework for analyzing how a problem such as discrimination against people with disabilities comes to be seen as a social problem. In his idea of the ownership of public problem, it is understood that all groups do not have the same power, influence and authority to define social problems. A group must truly own a problem to push it into the public sphere.

For example, disability organizations and disability activists had tried to "own" the problem of full civil rights for people with disabilities since the 1960s. In the mid-1970s, disability activists held sit-ins across the nation to protest the lack of enforcement guidelines of the Rehabilitation Act of 1973, which made discrimination against people with disabilities illegal at institutions that received federal money. However, that activism was mostly forgotten by the American public.

It was not until the late 1980s that the disability community truly owned the problem of discrimination against people with disabilities. Events such as the 1988 Deaf President Now student demonstration at Gallaudet University to protest the appointment of a hearing president at the university for deaf persons and national polls that delineated the problem of unemployment among people with disabilities gave the disability community more ownership of the discrimination problem. With that ownership, the disability community was able, through better definition of the problem, to transfer the responsibility for the problem to the U.S. government. Through the Americans with Disabilities Act, this public problem was fixed upon the whole of U.S. society, especially business concerns.

Gusfield explains that a component of this culture of public problems is mass media. Media help construct the "reality" of a public problem. In the case of the Americans with Disabilities Act, news media had only a little knowledge of the disability rights perspective at the inception of the Act, so they relied on the government and business perspectives they traditionally heard. This affected the message about disability that was disseminated in the ADA stories. In addition, Haller (1993) confirmed that journalists were just beginning to cover disability issues differently in the late 1980s because of the Deaf President Now protests. Johnson (1988) argues that the journalists saw these protests as "safe" disability stories, easily solved through the appointment of a deaf president; however, journalists had yet to understand the far-reaching implications of full civil rights for people with disabilities. Therefore, journalists came to the ADA story with only event-oriented knowledge of disability rights, so the messages from government and business easily colored the coverage.

Consequently, this affected the character of the news that the public received about disability rights. As Bird and Dardenne (1988) explain, the news that audiences receive is not just facts and figures but a larger symbolic system. Therefore, journalists transmit cultural narratives about disability, and they do so using cultural norms of how society views people with disabilities. Phillips explained the three themes about the disability experience that pervade in U.S. culture:

> (1) that society perceives disabled persons to be damaged, defective, and less socially marketable than non-disabled persons; (2) that society believes disabled persons must try harder to overcome obstacles in culture and should strive to achieve normality; (3) that society attributes to disabled persons a preference to be with their own kind (Phillips 1990, p. 850).

These cultural themes are likely to be imbedded in news stories on disability topics, and they depict a certain image of disability for the U.S. news audience. Phillips' themes are imbedded within the media models developed by Clogston (1989, 1990, 1991) and Haller (1993, 1995) to investigate the representation of people with disabilities in news stories. Clogston first created two categorizations of media portrayal of disabled persons: traditional and progressive. A traditional disability category presents a disabled person as malfunctioning in a medical or economic way. The progressive category views people as disabled by society, not a physical attribute.

Haller (1993) successfully applied Clogston's categories to a study of news media coverage of the deaf community. She looked at coverage of deaf persons before, during, and after the 1988 Gallaudet University tudent protest and found that the *Washington Post* and *New York Times* presented deaf persons in a more progressive and rights-oriented way after the protest. Gene Burd (1977) had shown in the 1970s that disabled persons and older persons seemed to receive the same type of media coverage that other minority groups such as poor people and African Americans received in the 1960s, thus strengthening the linkage between the disability minority group and other minority group representations.

The models used in Clogston's categories, which were developed before the advent of the Americans with Disabilities Act, were part of the assessment of the ADA coverage for this research, so as to understand how much validity they still have for this new type of story. However, three additional models (Haller 1995) were created for the study of the ADA stories to more accurately assess changing media narratives on disability rights issues.

The traditional categories include the medical model, the social pathology model, the supercrip model, and the business model.

In the medical model, (Clogston 1991) disability is presented as an illness or malfunction. Persons who are disabled are shown as dependent on health professionals for cures or maintenance. Disabled individuals are passive and do not participate in "regular" activities because of disability.

In the social pathology model, (Clogston 1991) people with disabilities are presented as disadvantaged and must look to the state or to society for economic support, which is considered a gift, not a right.

In the supercrip model, (Clogston 1991) the person with a disability is portrayed as deviant because of "superhuman" feats (an ocean sailing blind man) or as "special" because they live regular lives "in spite of" disability (a deaf high school student who plays softball). This role reinforces the idea that disabled people are deviant—that the person's accomplishments are amazing for someone who is less than complete.

In the business model, (Haller 1995) people with disabilities and their issues are presented as costly to society, and businesses especially. Making society accessible for disabled people is not really worth the cost and overburdens businesses. It is not a "good value" for society or businesses. Accessibility is not profitable.

Progressive categories include minority/civil rights model, legal model, cultural pluralism model, and consumer model.

In the minority/civil rights model, (Clogston 1991) people with disabilities are seen as members of a community that has legitimate political grievances. They have civil rights that they may fight for, just like other groups. Accessibility to society is a civil right.

In the legal model, (Haller 1995) it is illegal to treat disabled people in certain ways. They have legal rights and may need to sue to guarantee those rights. The Americans with Disabilities Act and other laws are presented as legal tools to halt discrimination.

In the cultural pluralism model, (Clogston 1991) those with disabilities are seen as a multifaceted people and their disabilities do not receive undue attention. They are portrayed as non-disabled people would be.

In the consumer model, (Haller 1995) people with disabilities are shown to represent an untapped consumer group. Making society accessible could be profitable to businesses and society in general. If disabled people have access to jobs, they will have more disposable income and willl no longer need government assistance.

In addition, Haller's use of the models as an assessment of the stories illustrates that several models may be apparent in one story, which was especially appropriate in ADA stories in which government, business, and disability rights sources came together. Past studies of disability images and issues confirm the validity of the themes covered by these models (Zola 1985, Knoll 1987, Biklen 1987, Cumberbatch and Negrine 1992, and Hevey 1992). Additional studies of journalism also showed that many times disability issues were misunderstood and therefore misrepresented by journalists (Gilbert 1976, Gardner and Radel 1978, Byrd 1979, Bonnstetter 1986, Yoshida, Wasilewski and Friedman 1990, Keller et al 1990).

METHODOLOGY

This study assessed the news media framing of the Americans with Disabilities Act. The sample consisted of all news and feature articles written about the ADA from 1988 through 1993 in the *New York Times, Wall Street Journal, Washington Post, Christian Science Monitor, Los Angeles Times, Chicago Tribune, Boston Globe, Atlanta Journal-Constitution, Philadelphia Inquirer, Newsweek, Time,* and *U.S. News and World Report.* These publications were chosen because they all are indexed and represent the major newspapers and news magazines of this country. They also represented four geographic regions in the country, the Northeast, the Midwest, the South, and the West, in addition to including a major business publication with the largest daily circulation, *The Wall Street Journal*.

It should be noted that the *Philadelphia Inquirer* represented a major portion of the sample. The newspaper's full library of stories from 1987 to present was available in database form. This is the entire newspaper for these years, so the sample includes all Neighbor Section stories, business briefs, and so on. The National Newspaper Index, which was used to find the stories in the other newspapers, only indexes stories and does not typically include briefs. In addition, all photographs appearing with ADA stories were analyzed with a separate coding scheme for photographs (N=171). The sampling procedure only picked up photographs that appeared with stories.

Within the newspapers and news magazines, one story on the ADA was the unit of analysis for coding purposes. For the visual communication sample, one photograph was a unit of analysis. The coding scheme was found statistically valid through an intercoder reliability test on 10 percent of the universe of stories.[1]

FINDINGS

The findings reported were based on a quantitative content analysis of 524 print news stories on the Americans with Disabilities Act from 1988 through 1993 in the elite news media. The types of stories split fairly evenly between hard news (45.6%) and feature stories (47.9%). Columns (3.8%), analyses (1.5%), and event notices (1.1%) accounted for the rest of the story types. Most of the print stories were found to be local (55%), meaning they focused on city or regional concerns, whereas 40.5% were found to be national stories and 4.6% focused on the impact to one state alone.

In addition, the analysis of 171 news photographs found that most of the photos were in the *Philadelphia Inquirer* (47.4%), followed by the *New York Times* (15.2%) and the *Chicago Tribune* (11.1%). A majority of the photos appeared on the front page of a section (60%), and the greatest percentage appeared in the neighbor or suburban section (36.8%), followed by the business section at 22.2%. Photos allowed for demographic information related to race, gender, and disability type to be revealed most accurately. It should be noted that one of the publications used in this study does not contain photos—the *Wall Street Journal*.

The placement of the ADA stories in different sections of the newspapers indicates how important the article is or the potential slant of the article. The ADA stories were most often in the Neighbors section at 25%, in the Front section 21% of the time and the Metro section 21% of the time. The ADA stories appeared in the business section of a newspaper 15% of the time.

Media Messages about the ADA

One way to assess how media assimilate new perspectives is to look for the reasons for change cited. In the stories about the Americans with Disabilities Act, the media quickly understood the civi l rights rhetoric in the stories, especially about architectural access and general discriminatory practices.

Table 1 illustrates that the news media represented architectural access as the most important reason for the ADA, followed by the issue of discrimination. Most stories (67.2%) did not cite a secondary reason for the ADA, but those that did came up with the same top four:

1) general inaccessibility of society (10.1%);
2) access to jobs (8.8%);

Table 1. Most Important Reason for the ADA as Cited in News Stories

	Frequency (N=524)	Percentage (100%)
Architectural access	139	26.5%
General discrimination against disabled people	116	22.1%
Access to jobs	96	18.3%
General inaccessibility of society	84	16.0%
Health insurance discrimination	14	2.7%
Health care concerns	10	1.9%
New technology	6	1.1%
Educational discrimination	4	.8%
Disability activism	2	.4%
Job loss	2	.4%
President pushed for it	2	.4%
No reason cited	49	9.4%

3) general discrimination against people with disabilities (5.7%);
4) architectural access (5.5%).

This study tried to assess how the news media perform when government, business and disability concerns intersect in a news story. Although the federal government controlled the timing of the ADA moving into public discourse, the news media went along with it. In the newspaper stories, 15.5% appeared in the month of July—the month the ADA was signed and that some of its major provisions took effect.

Another good way to understand how these three groups are reflected in news stories is to look at what sources are cited in the stories. The most prevalent sources clearly represent business, government, and disability groups. Significantly, people with disabilities (30.2%) or representatives of disability groups (35.3%) were represented as sources most often. Government spokespersons (31.3%), business people (30.2%), or business groups (25%) represented three of the top five source categories.

The sourcing illustrates that government interests and business interests are in a three-way competition with disability interests to characterize the ADA in media stories. But in this case, I would argue that the federal government's role in defining the law is more positive because disability activists helped write ADA. Also, it is federal legislation that was overwhelmingly passed in both the Senate and the House. State and local governments, however, may have a more negative relationship to the law; some local governments called it an "unfunded mandate" that would cause them economic hardship.

As expected, mentions of government groups dominated the ADA stories and were connected to the stories much more than disability groups were. Some type of government group was represented in 80.5% of the stories, compared to disability groups being represented 47.3% of the time. Reporters on the ADA print stories did seem to find their local disability organizations for comment, however, because local disability groups most often appeared in the stories at 19.3% of the time. It is significant to note that local government had strong representation in the ADA stories at 16.4%.

The Equal Employment Opportunities Commission (EEOC) (17.2%) and the Justice Department (12.2%) were two of the most often mentioned federal government groups because they are the two agencies charged with enforcing most of the ADA provisions in employment and accessibility. Interestingly, the ADA story was given cultural power through its linkage to the White House and president. The ADA stories mentioned the White House or president 13% of the time, although the president's advocacy was rarely used as a reason for the ADA (Table 1) and he was not a top source (2.3%).

A number of connections were made between government officials and a relationship to disability. In asking about a non-disabled source's family relationship to disability, about 5% of the stories showed the connection. Many of these sources were government officials who helped push for the ADA. Richard Thornburgh, attorney general at the time, has a son who was brain-injured in a car accident. President George Bush has one son with a learning disability and another with a colostomy. Sen. Lowell Weicker's son has Down Syndrome. Sen. Tom Harkin's brother is deaf. So government officials sometimes were used to validate the universality of the disability experience.

However, the news media failed to put the ADA in historical-governmental context. Only 5.7% of the articles mentioned the Rehabilitation Act of 1973, which mandated that any entity that receives federal funds not be allowed to discriminate against people with disabilities. The Rehab Act had been primarily ignored for decades so most government-related sources were not likely to mention it. Only 1.3% of the stories mentioned the other federal laws dealing with disability, such as the Fair Housing Standards Act. Only 6.5% of the articles mentioned local laws related to disability.

The newer media messages about disability in the ADA stories came from both its critics and supporters. The study investigated the evidence

Table 2. Prevalence of Models in ADA Print Stories

	Prevalent (N=524)	Evident (N=524)	Not Evident (N=524)
Minority/Civil Rights Model	27.1%	30.3%	42.6%
Business Model	15.5%	34.7%	49.8%
Legal Model	21.6%	27.7%	50.8%
Consumer Model	15.1%	22.3%	62.6%
Social Pathology Model	.4%	6.1%	93.5%
Medical Model	0	5.3%	94.7%
Supercrip Model	0	3.4%	96.6%
Cultural Pluralism Model	.2%	2.5%	97.3%

Notes: Several models may be represented in one story. "Prevalence" in the coding was defined as the model appearing as a dominant theme in the news article. "Evidence" was defined as the model appearing only slightly in the story, through a source, one instance of language, and so on.

of the eight media models in the ADA stories. They were treated as narrative themes that were not mutually exclusive, so several models may compete in one story. This allows for a more realistic assessment of the narrative conflicts within the stories. Table 2 illustrates the findings based on the representations of the models in the print stories.

In the ADA print stories, little evidence of the Medical Model was found, but examples did crop up. A story about a Philadelphia man who had 70% of his feet amputated but still was able to dance at his daughter's wedding had elements of the medical model because of its focus on him being a miracle of modern medicine (Samuel 1990, p. B1). As mentioned, most of the stories contained elements of the Minority/Civil Rights Model or the three models created for this study.

For example, a front page *New York Times* story quoted ADA supporters in the second paragraph, saying it is "the most sweeping civil rights bill in two decades" (Holmes 1990, p. A1). This story contained a prevalent minority/civil rights theme.

In contrast, a story with the Business Model focused on the Act's harmful effect on business. A story in the *Atlanta Journal-Constitution* was headlined, "Disabilities act closing popular meeting room" (Guevara-Castro 1992, p. D7), while a story in the *Chicago Tribune* was headlined "Suburbs face challenge, costs of satisfying disabilities act" (Thomas and Ostrowski 1992, p. NW3). Another Atlanta story said "Disabilities Act has high cost of compliance" (Snow 1993, p. M1).

The legal model was not difficult to find in stories once the law took effect because people began using the ADA to confront discrimination. The *Washington Post* gave front page play to the first Justice Department suit under the ADA. The story reported how two police officers had been denied retirement, disability pensions, and survivors' benefits because one of the men has diabetes and one has chronic back problems (Duke 1993, p. A1). An earlier story illustrated the legal model when it reported on the trial expected because the U.S. Department of Health and Human Services had denied a woman with multiple sclerosis the right to work at home (Spayd 1992, p. A25).

Another *Washington Post* article embodied the consumer model when it began its lead with: "Lee Page's money is as green as anyone else's, but he regularly encounters difficulty spending it in many Washington businesses. Frugality isn't the problem. Physical barriers are" (Lehman 1990, p. E1). Also in line with the consumer model, the *Wall Street Journal* regularly reported on the growth of disability-related business opportunities such as software entrepreneurs making money by creating programs that will assist businesses in ADA compliance (1992, p. A1).

The strong evidence of the business model in the ADA stories illustrated the conflicting message that the news media disseminated about disability issues: People with disabilities deserved equal access, yet businesses should not be burdened with the cost of accommodations. This conflict continues in some of the backlash against the ADA rhetoric that surfaced in the mid- to late 1990s. From the first proposal of the ADA, many businesspeople said that it would be too expensive for businesses and other sectors of society to implement. To confirm this link between the "bad for business" rhetoric and business and some government sources, cross-tabulations were run on mentions of the costliness by sources and groups. In terms of percentages, this notion of costliness associated with the ADA was found more often in stories with sources from business or mentions of some government groups.

Table 3 illustrates the trend toward mentions of costliness by selected sources and government and disability groups. Most of the cross-tabulations for this table were not statistically significant; however, it does

Table 3. Mention of Costliness of ADA by Selected Sources and Groups

	Number of Occurances	Percentage	Number mentioning high cost of ADA	Percentage mentioning high cost of ADA
Totals	524	100%		100%
Representative of business	105	20%	33	31.4%
Business person	131	25%	38	29%
Government agency spokesperson	164	31.3%	38	23.2%
Representative of a disability group	185	35.3%	39	21.1%
Person with a disability	158	30.2%	21	13.3%
City government	86	16.4%	34	39.5%
EEOC	90	17.2%	23	25.6%
White House/ President	68	13%	17	25%
Local disability group	101	19.3%	14	13.9%

show a crucial trend in the "bad for business" message in the ADA stories. The reporting trend this illustrates seems to be that if disability organizations and activists strongly countered the "bad for business" message as sources in the news, a less stigmatizing story would result.

For example, Joe Shapiro, a reporter for *U.S. News and World Report* and author of a well-regarded book on disability rights, illustrated this phenomenon in his reporting on the Americans with Disabilities Act. A story of his addresses the probable cost of accommodations under the ADA and quotes complaints about the expense from business groups, but the story is balanced with comments from Ron Mace of Barrier Free Environments, who explained that most accommodations will be "simple and cheap" and from Lex Friedan of the Institute for Rehabilitation and Research Foundation, who said "We don't want to be dependent any more" (Shapiro 1989). A number of other prominent disability sources are quoted as well, giving realistic information about the ADA's impact on society, thus countering the "bad for business" narrative.

Representations of People with Disabilities in ADA Stories

This study also looked for specifics based on race, gender and disability type. It should be noted that almost half of the ADA print stories had no person with a disability as a source or example (49.2%). About 44% of the stories mentioned an organized disability group. In the photographs, 87.7% of the people with disabilities were adults, 6.4% were teenagers, and 4.1% were children. In the print stories, 23.5% of the people with disabilities were men and 10.1% were women. Both men and women with disabilities were represented in 14.7% of the stories. The news photos depicted 68 men and 55 women alone. Nineteen photos had both men and women shown.

The 10 most prevalent disabilities mentioned in the print stories were wheelchair use (48.3%), deafness (23.5%), blindness (22.7), AIDS (14.3%), paralysis (9%), cerebral palsy 8%), major mental illness (6.9%), cancer (5.3%), mental retardation (4.8%) and polio/post polio (4.8%). By comparing these findings to the actual incidence of disability in the United States, one can see that the stories and pictures give a somewhat distorted view of the prevalence of certain disabilities.

The top 10 disabilities in the United States that could fit within the ADA definition[2] are: arthritis, affecting 30 million people; mental disorder (excluding substance abuse) which affects 23.9 million people; hearing impairments, which affect 23 million people; heart conditions, which affect 19.3 million people; cancer, of which 8 million have a history; visual impairments, which affect 7.5 million people; mental retardation, which affects 6 million people; diabetes, which affects 6.2 million; and paralysis (full or partial), which affects 1.5 million people (Center for Health Statistics 1990, President's Committee 1994).

This comparison shows that the visible and more severe disabilities received the coverage. Wheelchair use appears to be the symbolic disability for the news stories, although its actual incidence in the U.S. population is 1.5 million out of about 48 million people with disabilities. Obviously, the architectural access issues surrounding wheelchair use gives it somewhat greater weight in society. But people who use canes, crutches, walkers, artificial legs or who are blind have similar access concerns, yet they do not seem to have the symbolic value that wheelchair use does.

As the ADA was implemented to alleviate employment discrimination, the actual incidence of disability rather than the media version was also confirmed. Of the 24,730 charges filed through the EEOC from July 1992 to March 1994, the largest percentage were by people with back impairments (20%), neurological impairments (13%), and emotional/psychiatric impairments (11%) (EEOC 1994). Several of the first cases filed under the ADA involved people with cancer, diabetes, visual impairments and hearing impairments

Because race was not mentioned in the print stories, the best assessment of race and gender represented in the news media came from the analysis of the news photos. Table 4 illustrates those results.

Table 4. Gender and Race of People with Disabilities Pictured in News Photos

	White	Black	Asian	Latino	Mixture of Ethnicities	% by Gender[a]
Totals	123	17	1	3	19	
Male	68	8	0	2	0	46.2%
Female	55	9	1	1	0	40.4%
Mix	0	0	0	0	19	11.1%
Percentage by Race	80.1%	10.5%	.6%	2.3%		

Notes: Chi Square = 112.07. Significance = .000.
[a] Eight of the photos were poor reproductions in which it was difficult to tell the race or gender of the photo subject. These represent missing values of 2.3% of the photos by gender and 4.7% of the photos by race.

When Tables 4 is compared to other demographic data, it illustrates that news stories also somewhat misrepresented the race and gender of people with disabilities. Men with disabilities show up most prominently in the news stories (23% male, 10% female) and photographs, while according to the statistical studies women (53%) more often have disabilities than men (47%) (President's Committee 1994).

The racial demographics are more debatable, but many disability researchers believe the incidence of disability in ethnic groups has been underestimated in the past. A 1991 study by the Disability Statistics Program at the University of California-San Francisco, reports that 27.5% of whites have some form of disability, 34.4% of African Americans, 26.8% of people of Spanish origin, and 20.1% of people of other races.

With heavy reliance on pictures of white people with disabilities, the newspapers construct disability as a "white issue," which, of course, it is not. The image that begins to appear from the news stories is one of a white male wheelchair user trying to get into a restaurant or office. Less mention is made of the discrimination he might face in the workplace or the fact that many African American, Asian, or Latino men and women with disabilities have not made it into the workplace.

In addition, demographic information about the metropolitan areas where the 12 publications in this study are based confirmed that these publications reside in areas with high racial diversity. For example, the New York area is 18% African American, as is the Philadelphia area. Atlanta is 25% African American; Chicago is 19%, and Los Angeles is 9% Asian (U.S. Census 1993). All these percentages are significantly higher than the national numbers for these ethnic groups. The findings of past studies on race and gender in news photos mirror the findings of photos with ADA stories—many more men and whites are typically depicted (Miller 1975, Singletary 1978, Blackwood 1983).

DISCUSSION

The coverage of the Americans with Disabilities Act illustrated that the notion of disability rights is only making a moderate amount of headway in news media representations. The coverage was spotty and sometimes shallow and reflected almost no historical context of changes in the treatment of people with disabilities. One might argue that this results because the ADA was not a big story compared to health care reform, the Persian Gulf War or Iran-Contra. But that is untrue: The ADA was and is a big story because it attempts to change the fabric of U.S. society. In addition to affecting 43 million to 48 million U.S. citizens, it legislated how the country organizes its workplaces, how it provides access to most of the architectural structures in the country, how its citizens travel and communicate.

Yet only 288 stories in 12 prestigious newspapers and news magazines over five years focused primarily on the ADA. Network television news went for more than a year several times without mentioning the Act (Haller 1995). This neglect probably had a real impact on public understanding of the legislation because in 1991, only 18% of Americans said they had heard of the ADA (Harris 1991). Therefore, the news media agenda-setting function in this coverage was to ignore the sweeping societal ramifications of the ADA and portray it as another piece of government legislation that would only affect people with significant disabilities and businesses.

However, the findings in this study are unique in that they illustrate that the news media's reliance on government sources can work in the favor of a marginalized group, if that group had a hand in creating federal legislation. It was fortunate for the disability rights movement that Congress did not dilute their perspective in the ADA because that allowed the media to latch onto something validated by a major government institution.

The legislative nature of the ADA story also did not allow the media to use some of the traditional stereotypes of the past, which presented people with disabilities as medical problems in need of a cure or as superhuman (Clogston 1990). This story forced the news media to acknowledge people with disabilities as having minority-group status and deserving full civil rights. It seems clear that journalists can assimilate new frames, especially when they are handed down from Congress. This is significant when that new frame tries to shake loose old stereotypes.

Watson said the greatest accomplishment of the disability community in getting the ADA passed was maintaining the narrative of civil rights and minority group politics in the law. The disability community established the bipartisan nature of the legislation and then established the reason for the ADA—"that its protections were an issue of civil rights rather than a charitable obligation or some other rationale." The narrative they hung onto most strongly, therefore, was "civil rights regardless of cost" (Watson 1993, pp. 29-30). With that narrative secure in the legislative language, government sources and disability-related sources gave the news media the same information.

A secondary rationale for the ADA used early on by the disability community was cost effectiveness: Society is better off if people with disabilities become taxpayers instead of tax beneficiaries. This was referred to as the consumer model of representation in the study. The narrative of cost effectiveness took on a lesser role in the media framing because Watson (1993) says the activists dropped this argument when they realized some accommodations under the Act would be expensive.

For decades, disability has been defined and framed by government through legislation on war veterans, rehabilitation, education, and Social Security, but this time the disability rights movement had rhetorical power to craft the ADA. As Scotch explains: "The disability rights movement is one in which the way an issue was framed had serious effects on both movement participation and the ability of the movement to influence public policies" (Scotch 1988, p. 168). The federal government finally accepted the disability rights frame.

However, in sticking to the norms of their adversarial role, journalists did provide contrast to the government's rhetoric by going to business and local government as sources in the news stories. The business community, fearing the financial ramifications of the ADA, supplied information to the media for a new frame—that the ADA would be costly to business. It is not surprising that the news media also would embrace business sources, considering media are businesses that also have to comply with the ADA. Shoemaker and Reese (1991) say media messages fit within the basic ideology for the United States:

> Fundamental is a belief in the value of the capitalistic economic system, private ownership, pursuit of profit by self-interested entrepreneurs, and free markets. This system is intertwined with the Protestant ethic and the value of individual achievement (Shoemaker and Reese 1991, p. 184).

This ideology was perpetuated and embodied by the business sources found in the ADA stories. Even the reasons cited for the creation of the Americans with Disabilities Act illustrate the credence given to the "business" side of the story. Although the ADA is a civil rights bill dealing with myriad issues, especially employment discrimination against people with disabilities, the issue of architectural access was cited most often as the reason for the ADA. In addition to being the visible result of the Act, architectural access also has the potential to be most costly to business and government.

However, the media's reliance on government, disability and business sources may have shut out other kinds of information that were significant to the ADA or stories about disability issues. One frame that should have arisen in 1992-1993 was whether the ADA was accomplishing its purpose. But the news media followed their norm of event-driven journalism, and only rarely stepped into their watchdog role to scrutinize the impact and enforcement of the ADA. That scrutiny usually came from a lawsuit-related story on someone suing for access or workplace accommodation under the ADA, which would have fallen into the legal model of coverage.

Mostly, news media shirked their investigative role to assess the ADA. There were a few exceptions, however. *The Wall Street Journal,* for example, wrote a 1993 story headlined: "Disabilities act helps—but not much. Disabled people aren't getting more job offers" (Quintanilla 1993, p. B1). The story explained that more people with disabilities had yet to move into the workplace because of the ADA. And a *Washington Post* story explained how a school guidance counselor with multiple sclerosis in New York had been trying to use the ADA to receive workplace accommodation with no success (Mathews 1993).

But these types of scrutinizing stories were few and far between. Once again, as Olien, Tichenor and Donohue (1989) have said, the news media became "lap dogs" for mainstream interests, in this case the government that was not quickly and effectively administering the ADA, rather than "watchdogs" for the interests of all people affected by the ADA.

The Backlash Issue

The media frames from the Americans with Disabilities Act coverage continue to contribute somewhat to the current anti-ADA sentiment that can be heard in some segments of society in the late 1990s. Shapiro (1993) has always argued disability rights activists made the wrong decision to exclude the media from their lobbying efforts regarding the ADA. They feared the media would just perpetuate the same old stereotypes, so lobbyists avoided discussing the ADA with journalists, according to Shapiro.

However, the findings from this content analysis of ADA coverage illustrate the depth of that miscalculation. Activists did not understand the nature of the media-government relationship, which according to popular myth is adversarial, but in actuality is quite congenial and unquestioning. The ADA coverage illustrated how accepting the news media are of bipartisan legislation and civil rights movements, echoing the disability rights narrative across the nation in complete contrast to their past stereotypical images of disability. Basically, this coverage potentially opened the door to nonstigmatizing messages about people with disabilities, but instead of disability activists stepping in to continue a process of change in media coverage, the business community did.

For example, with the media so readily embracing the disability rights rhetoric in the governmental language of the ADA, activists could have loudly trumpeted the secondary rationale for the ADA: Society is better off if people with disabilities become taxpayers instead of tax beneficiaries. Therefore, the media coverage would have had a more equal counternarrative to the business community's "bad for business" rhetoric. However, activists dropped the cost-effectiveness argument because they realized some accommodations under the ADA could be expensive (Watson, 1993). This left the news coverage with only one side of the "costliness" story—the business community's.

Therefore, part of the ADA's message was lost: That it took into consideration many of the concerns of businesses without gutting the intent of the law. The Act only covers businesses with more than 15 employees. Employment studies show that only 22 percent of people with disabilities need accommodations at the work site. Another study showed that 50 percent of all accommodations cost $50 or less (Eastern Paralyzed Veterans Association 1992). This is the kind of information disability activists should have been feeding the media continuously to counter the business community's rhetoric.

The image problems caused by the business model's prevalence in the ADA stories continues today. In November 1998, nationally syndicated *Washington Post* columnist William Raspberry complained that a blind man asking for the Bay Area Rapid Transit's (BART) web site for bus and train schedules to be accessible under the ADA was "a clear violation of common sense" (Raspberry 1998, p. A9). Raspberry claims to support the ADA, in terms of ramps and curb cuts, but he says he is angry with people with disabilities who insist "their disability be accommodated to and that we take no notice of it." He goes on to complain about buses with chair lifts and people who must use telecommunications devices for the deaf (TDDs) when others are pressed for time.

In addition, his business model argument ignores the civil rights narrative so prominent in early media coverage of the Act. Raspberry is African American, writes about race, poverty and other social issues, and won a Pulitzer Prize for distinguished commentary in 1994. However, this column fails to show any understanding of the civil rights movement of people with disabilities, which was informed by the 1960s Civil Rights Act of which Raspberry was a beneficiary.

In responding to Raspberry, Randy Tamez, the blind man who is suing BART, confronts Raspberry's anti-disability and segregationist rhetoric with a civil rights narrative and reminds us that "it is obvious that the biggest obstacle to access is the attitudinal barrier" (Tamez 1998). However, these backlash attitudes will have the best chance of changing if the consumer narrative is used in conjunction with the civil rights narrative. For example, making the web site accessible for Tamez by creating a text-based site is very low cost and allows him to use BART, rather than a more expensive "special" disability transportation system.

In terms of representation of people with disabilities, the ADA stories, as disability rights lobbyists feared, got many disability demographics wrong. The media sought out the visible disabilities as examples and missed the fact that more people have hidden disabilities. They portrayed disability in terms of the white middle class. This may reflect the racial and class dynamics of media workers, but is not necessarily representative of their audiences or the U.S. population at large.

For example, a lobbyist for the Paralyzed Veterans Administration was pictured on a ramp to a D.C. sandwich shop (Herndon 1990), a psychotherapist sat before his accessible van (Barnard 1992), or a New Jersey disability activist/community leader used the wheelchair entrance at a train stop (Gekoski 1992). In a story in the *Los Angeles Times*, a Latina woman with a disability was quoted extensively in the article, yet the photos were of a white male wheelchair user who was not even a source in the story (Newton, 1990).

Here was another place in the coverage where disability rights lobbyists could have turned the norms of journalism in their favor. They could have easily supplied news media with the demographic truth of disability—most people who have disabilities are women, many have "invisible" disabilities, a large number of the disability community belongs to an ethnic minority group and a significant number of people with disabilities are unemployed. In fact, in 1990 an estimated 8.2 million working-age adults with disabilities wanted to work but were unemployed (National Organization on Disability 1991).

Instead, a white, male wheelchair user became the "representative" person with a disability in the news media. Visually, wheelchairs give journalists and their audiences cues on which to hang the story. In print stories especially, they allow the mind's eye to "see" the person in the story even if no photos accompany the article.

In addition, the focus of disability rights movement on architectural access was probably another reason mobility impairment was used by the media to illustrate disability. Even if wheelchair users had less impairment than other types of disabilities, they still could not enter the grocery store, post office, garment factory, etc., if these places are not accessible. Altman explained it this way:

> Visibility becomes paramount in differentiating the "oppressed" from the oppressors. So, during the civil rights movement we saw black leaders with much darker skins than were generally the case prior to the civil rights movement, when light skin was preferred. In the same way, persons who used wheelchairs became the most effective protestors during the early years of the disability rights movement because of their visibility and because their physical access problems were also visible (Altman 1994, p. 49).

This visibility was used to advocate for rights and thus became imbedded within news coverage.

The implications of these misrepresentations that flow through the news media are that they contribute to an on-going backlash and misunderstanding of disability rights. The fact that the news media portrayed disability as primarily Caucasian and male means that the news media assisted in perpetuating the image of a dominant societal group. In a public relations sense, it may be favorable to ethnic minority groups to be portrayed as disability free, especially considering negative cultural notions about disability. But actually perpetuating this frame can have dire consequences to groups already oppressed by the majority. The National Council on Disability (1993) reports that because of poverty, unemployment and poorer health, many members of minority groups may be at higher risk of disability.

This type of monolithic representation also potentially clouds public understanding of discriminatory practices, rather than illuminating them. For example, in the ADA coverage only 3.6% of the stories mentioned racial discrimination. The ADA was built upon the anti-discrimination foundation of 1964 Civil Rights Act, yet these connections were not made. As Shapiro (1993) explains, the news media were accepting of the 1964 Act and even instrumental in public understanding of it. But the "stealth civil rights movement" of disability lobbyists resulted in an important disability rights law such as the ADA being passed easily, but with little public or media understanding of it.

The paradigm has shifted governmentally to a disability rights perspective because of the Americans with Disabilities Act, but the media coverage illustrates that it has yet to completely shift culturally. Shapiro's prediction of a backlash because of this has come partially true. And this backlash is embodied in the continuing narrative from the business community and the possible weakening of the consumer model in media stories.

However, some current media stories give some hope that the news media learned something about disability issues in their coverage of the ADA. In the Tamez case, several media stories clearly understand the importance of the ADA covering web communication. In an Associated Press story, there is no mention of the cost of making web sites accessible, only discussion of how bad online design can create barriers to people with disabilities (AP 1998, Nov. 12). Because of the Tamez case, Fox News Online published a story on web accessibility that fell easily into the consumer model with its detailed discussion of the number of blind people who own computers and use the Internet and other online services (Riley 1998, Nov. 20).

Accurate media messages such as these have positive ramifications for people with disabilities and society in general. Cultural representations create important definitions of what it means to be physically different and this can have consequences for future social policy about disability issues (Berube 1997). For example, if the news media associates disability with a negative such as societal costliness, rather than with a positive such as more employed taxpayers, it could have dire consequences for people with disabilities taking their rightful place in the U.S. economic structure. By accurately representing the reality of the disability experience in America, news media can help the public accept disability as just a variation of the human experience, not as a tragedy or burden on society.

NOTES

1. To assess the reliability of the coding scheme, 50 stories (about 10 percent) were coded by another coder. The findings on the majority of the variables within the code sheet were compared between the author of this study and the coder. Most of the code sheet questions were involved in the intercoder reliability test. The answers to these questions were then collapsed into one mean for the author's answers and one mean for the coder's answers. A t-test was run, and it illustrates the similarities between the author's findings and those of the coder. The author's mean was 4.50, standard deviation = 13.45 and the standard error =.28. The coder's mean was 4.56, standard deviation =13.32 and standard error =.28. The F value from this t-test was 1.02 and the two-tailed probability was .619.

2. "An individual is considered to have a 'disability' if s/he has a physical or mental impairment that substantially limits one or more major life activities, has a record of such impairment, or is regarded as having such an impairment." (EEOC 1992, p. 1). People who have an association or relationship with someone who has a disability are also protected against discrimination. In this definition, "major life activities" include seeing, hearing, speaking, walking, breathing, learning, self care, working and performing manual tasks. Temporary or nonchronic impairments such as broken bones are not defined as disabilities. The concept of a "record of impairment" helps protect people who have recovered from an impairment, such as cancer or mental illness, from discrimination. The idea of "regarded as having" an impairment protects people who may not have an impairment that limits them physically, such as people with facial disfigurement or who are HIV positive but may face employment discrimination because of fear.

REFERENCES

Altman, B.M. 1994. "Thoughts on Visibility, Hierarchies, Politics & Legitimacy." *Disability Studies Quarterly* 14(2): 48-51.

Associated Press 1998. "Web focal to blind man's complaint." *USA Today Tech Report* (November 12): www.usatoday.com/life/cyber/tech/ctd831.htm.

Barnard, T. 1992. "Psychotherapist Thomas W. Fritz Hopes." (Photograph) *Los Angeles Times* (January 26): D1.

Berube, M. 1997. "The Cultural Representation of People with Disabilities Affects Us All." *Chronicle of Higher Education* 43(38): B4-5.

Biklen, D. 1987. "Framed: Print Journalism's Treatment of Disability Issues." In *Images of the Disabled, Disabling Images*, edited by A. Gartner and T. Joe. New York: Praeger.

Bird, S.E., and R.W. Dardenne. 1988. "Myth, Chronicle, and Story." Pp. 67-86 in *Media, Myths, and Narratives*, edited by J.W. Carey. Newbury Park, CA: Sage.

Blackwood, R.E. 1983. "The Content of News Photos: Roles Portrayed by Men and Women." *Journalism Quarterly* 60: 710-714.

Bonnstetter, C.M. 1986. "Magazine Coverage of the Mentally Handicapped. *Journalism Quarterly* X: 623-626.

Burd, G. 1977. "Aged and Handicapped Seek Human Quality and Public Service in the Media: Mass Communications Patterns of the New Minorities." Paper presented at the Annual Meeting of the Association for Education in Journalism, Madison, WI.

Byrd, E.K. 1979. "Magazine Articles and Disability." *American Rehabilitation*, 4(4): 18-20.

Clogston, J.S. 1989. "A Theoretical Framework for Studying Media Portrayal of Persons with Disabilities." Paper presented at the Annual Meeting of the Association for the Education in Journalism and Mass Communication, Washington, DC.

_____. 1990. *Disability Coverage in 16 Newspapers*. Louisville: Advocado Press.

_____. 1991. *Reporters' Attitudes Toward and Newspaper Coverage of Persons with Disabilities*. Unpublished doctoral dissertation, Michigan State University.

Cumberbatch, G., and R. Negrine. 1992. *Images of Disability on Television*. London: Routledge.

DeJong, G. 1993. "The Americans with Disabilities Act: A Study in Bipartisan and Consensus Development." Paper presented at the annual meeting of the Society for Disability Studies, Seattle, WA.

Dines, G. 1992. "Capitalism's Pitchmen: The Media Sells a Business Agenda. *Dollars and Sense* 176: 18-20.

Disability Statistics Program. 1991. "People with Functional Limitations in the U.S." *Disability statistics abstracts*, No. 1, Jan..

Duke, L. 1993. "Justice Department Sues Illinois, City Under Disabilities Act. *Washington Post* (December 29): A1, 6.

Eastern Paralyzed Veterans Association. 1992. *Understanding the Americans with Disabilities Act*, Jackson Heights, NY: Eastern Paralyzed Veterans Association.

Fasman, Z. 1989. "Should the Senate Approve the Americans with Disabilities Act of 1989?" *Congressional Digest* December: 299, 301.

Gans, H. 1980. *Deciding What's News: A Study of CBS Evening News, NBC Nightly News, Newsweek and Time*. New York: Vintage.

Gardner, J.M., and M.S. Radel. 1978. "Portrait of the Disabled in the Media." *Journal of Community Psychology* 6(3): 269-274.

Gekoski, S. 1992. "Think this looks easy?" (Photograph). *Philadelphia Inquirer* (January 23): C4.

Gilbert, L.J. 1976. *Media Understanding of Deafness*. Unpublished masters thesis. American University, Washington, DC.

Green, M.C. 1989. "Should the Senate Approve the Americans with Disabilities Act of 1989?" *Congressional Digest* December: 309, 311, 313.

Greyhound Lines Inc. 1989. "Should the Senate Approve the Americans with Disabilities Act of 1989? *Congressional Digest* December: 301, 303, 305.

Guevara-Castro, L. 1992. "Disabilities Act Closing Popular Meeting Room." *Atlanta Constitution* (May 7): XE7.

Gusfield, J. 1981. *The Culture of Public Problems*. Chicago: University of Chicago Press.

Haller, B. 1995. *Disability Rights on the Public Agenda: News Media Coverage of the Americans with Disabilities Act*. Unpublished doctoral dissertation. Temple University, Philadelphia, PA.

___. 1993. "Paternalism and Protest; The Presentation of Deaf Persons in the *Washington Post* and *New York Times*, 1986-1990." *Mass Comm Review* 20(3/4): 169-179.

Herndon, C. 1990. "Lee Page of the Paralyzed Veterans Association Uses a Ramp" (Photograph). *Washington Post* (December 29): E1.

Hevey, D. 1992. *The Creatures Time Forgot: Photography and Disability Imagery.* London: Routledge.

Higgins, P.C. 1992. *Making Disability: Exploring the Social Transformation of Human Variation.* Springfield, IL: Charles C. Thomas.

Holmes. S. 1990. "House Approves Bill Establishing Broad Rights for Disabled People." *New York Times* (May 23): A1, 18.

Johnson, M. 1988. "The Gallaudet Difference." *Columbia Journalism Review* (May/June): 48.

Keller, C.E., D.P. Hallahan, E.A. McShane, E.P. Crowley, and B.J. Blandford, 1990. "The Coverage of Persons with Disabilities in American Newspapers." *Journal of Special Education* 24(3): 271-282.

Knoll, J.A. 1987. *Through a Glass, Darkly: The Photographic Image of People with a Disability.* Unpublished doctoral dissertation. Syracuse University.

Lehman, H.J. 1990. "Disabled Hoping New Law Will Improve Building Access." *Washington Post* (December 29): pp. E1, E4.

Liachowitz, C. 1988. *Disability as Social Construct.* Philadelphia, PA: University of Pennsylvania Press.

Linsky, M. 1986. *Impact. How the Press Affects Federal Policymaking.* New York: W.W. Norton & Co.

Louis Harris and Associates, Inc. 1991. *Public Attitudes Toward People with Disabilities.* National poll conducted for National Organization on Disability. New York, NY.

Mathews, J. 1993. "Having Doubts about Disabilities Act." *Washington Post* (December 6): p. A21.

McCombs, M. 1992. "Explorers and Surveyors: Expanding Strategies for Agenda-Setting Research." *Journalism Quarterly*, 69(4): 813-824.

McCombs, M., and D. Shaw. 1972. "The Agenda-setting Function of the Press. *Public Opinion Quarterly*, 36: 176-187.

Miller, S. 1975. "The Content of News Photos: Women's and Men's Roles. *Journalism Quarterly* 52(1): 70-75.

National Council on Disability. 1993. *Meeting the Unique Needs of Minorities with Disabilities.* Washington, DC.

National Organization on Disability. 1991. Information brochure. Washington, DC.

Newton, E. 1990. "Disabled: The Battle Goes On." *Los Angeles Times* (August 15): pp. 1, 16.

Olien, C., P. Tichenor, and G. Donohue. 1989. "Media and Protest." In *Monographs in Environmental Education and Environmental Studies,* edited by L. Grunig. Troy, Ohio: North American Association for Environmental Education.

Phillips, M.J. 1990. "Damaged Goods: The Oral Narratives of the Experience of Disability in American Culture." *Social Science & Medicine*, 30(8): 849-857.

President's Committee on the Employment of People with Disabilities. 1994. *Statistical Report: The Status of People with Disabilities.* Washington, DC: Department of Labor.

Quintanilla, C. 1993. "Disabilities act helps—but not much." *Wall Street Journal* (July 19): B1, 2.

Raspberry, W. 1998. "Sometimes, Accommodations for Disabled Defy Common Sense." *The Washington Post*, (November 17): A9.

Riley, P. 1998. "Case Could Mandate Web-site Accessiblity for the Disabled." (November 20): Fox News Online, http://www.foxnews.com/

Samuel, T. 1990. "Better Living for the Disabled." *Philadelphia Inquirer* (August 27): B1, 2.

Scotch, R.K. 1988. "Disability as the Basis for a Social Movement: Advocacy and Politics of Definition." *Journal of Social Issues* 44(1): 159-172.

Shapiro, J. 1993. "Disability Policy and the Media: A Stealth Civil Rights Movement Bypasses the Press and Defies Conventional Wisdom." In *Disability Policy as an Emerging Field of Mainstream Public Policy Research and Pedagogy*, edited by S. Watson and D. Pfeiffer, a symposium edition of the *Policy Studies Journal* 22(1): 123-132.

_____. 1989. "Liberation Day for the Disabled." *U.S. News and World Report* (September 18): 20-23.

Shoemaker, P.J., and S.D. Reese. 1991. *Mediating the Message.* New York: Longman.

Singletary, M.W. 1978. "New Photographs: A Content Analysis, 1936-1976." *Journalism Quarterly*, 55(3): 585-589.

Snow, W. 1993. "Disabilities Act has High Cost of Compliance." *Atlanta Constitution* (October 12): XJM1, 14.

"Software Entrepreneurs Find a Lucrative Niche in the Disability Law." *Wall Street Journal* (June 25, 1992): p. A1.

Spayd, L. 1992. "Trial is Ordered in Denial of Work-at-home Request." *Washington Post* (April 2): A25.

Tamez, R. 1998. "Response to Editorial." http://lists.w3.org/Archives/Public/w3c-wai-ig/1998OctDec/0318.html.

Thomas, R., and S. Ostrowski. 1992. "Suburbs Face Challenge, Costs of Satisfying Disabilities Act." *Chicago Tribune* (July 15): NW2.

U.S. Bureau of the Census. 1993. *Statistical Abstracts of the United States: 1993*, 113th edition. Washington, DC..

U.S. Equal Employment Opportunity Commission. 1992. *The Americans with Disabilities Act: Questions and Answers.* EEOC-BK-15. Washington, DC.

_____. 1994. "Total number of ADA charges received by EEOC." Information release from EEOC office of communications and legislative affairs. Washington, DC.

U.S. National Center for Health Statistics. 1990. *Advance Data from Vital and Health Statistics.* No. 217. Washington, D.C.

_____. 1990. *Vital and Health Statistics.* Series 10, No. 181. Washington, D.C.

Watson, S.D. 1993. "A Study in Legislative Strategy." *Implementing the Americans with Disabilities Act.* Baltimore, MD: Paul H. Brookes.

Yoshida, R.K., L. Wasilewski, and D.L. Friedman. 1990. "Recent Newspaper Coverage About Persons with Disabilities." *Exceptional Children* 56(5): 418-425.

Zola, I.K. 1985. "Depictions of Disability—Metaphor, Message, and Medium in the Media: A Research and Political Agenda." *Social Science Journal* 22(4): 5-17.

GENDER CONTRADICTIONS AND STATUS DILEMMAS IN DISABILITY

Judith Lorber

ABSTRACT

Both women and men have benefited from the successes of the disability rights movement, but the special needs of women with disabilities for jobs, sexual relationships, and a family life have not been so squarely faced. In this paper, I discuss gender differences in the roles and social status of people with disabilities. I argue that there are contradictions and status dilemmas in the beliefs about the characteristics of women and men with disabilities and that gender-related behavioral expectations create constraints for both. I also discuss the expectation that caregivers will be women and the effect of this expectation on the social roles of women and men with disabilities. Melding gender norms would be helpful, but just as people with disabilities have special needs if they are to live mainstream lives, women with disabilities also have special needs for services, especially around sexuality and procreation.

Whoever I was, whatever I had, there was always a sense that I should be grateful to someone for allowing it to happen, for like women, I, a handicapped person, was perceived as dependent on someone else's largesse for my happiness, or on someone else to let me achieve it for myself (Zola 1982a, p. 213).

GENDER CONTRADICTIONS AND STATUS DILEMMAS IN DISABILITY

Within the last decade, the social status of people with disabilities has changed in ways reflected in that very designation—they are people first, not "the handicapped" or "the disabled," which erases their personhood. Yet the designation of "people with disabilities" embeds a *status dilemma*. When people have two contradictory major statuses, one high and one low, they can't establish their rightful place in their social world (Hughes 1971). People with disabilities have been able to assert their competence and capabilities in their work and family lives, but their status gets undercut by the negative attributes still attached to physical disabilities.

This status dilemma has often been handled by keeping coping strategies and physical help under cover—what sociologists call "deviance disavowal" (Davis 1972). For example, Bob Dole, who has little control over his right arm, always clutches a pen in his right hand to keep the fingers from splaying. When he campaigned for the presidency of the United States in 1996, his right side was protected from crowds by his aides, and they unobtrusively gave him a pad to lean on when he signed autographs (Kelly 1996).

In such encounters among people of different physical capabilities, interaction proceeds as if everyone were on the same footing. But the person who does not have the full use of limbs, eyes, ears, voice, or other bodily functions knows how much effort goes into making it possible to participate socially as an equal. Hiding that effort has made it seem somehow shameful.

Everyone cannot maintain the same level of physical exertion at all times. Even common conditions such as cardiac or breathing problems, difficulty with walking, or advanced pregnancy often need to be compensated for. Yet people with disabilities are not seen as part of a continuum but as qualitatively different. People who are seen as qualitatively different are often stigmatized. Stigma depends more on what can be seen than on what can be done: "The person who looks relatively 'normal' but is severely limited by what she can do yields a dif

Gender Contradictions and Status Dilemmas in Disability 87

ferent cultural figure from the person who performs life activities successfully—often with aids—but who looks 'abnormal.'" (Thomson 1994, p. 590).[1] Either way, people with disabilities have a status dilemma: In the eyes of able-bodied people, physical incapacities devalorize whatever else they accomplish.[2]

In order to maintain a high status, disabilities have had to be rendered invisible or transformed into heroism. Franklin Delano Roosevelt, a polio victim who served as president of the United States from 1932 to 1944, masked his inability to walk or to stand without supports (Gallagher 1985). John Hockenberry, a paraplegic due to an automobile accident, has gone around the world as a reporter in his wheelchair, admittedly flaunting his physical state (Hockenberry 1995).

What has changed so dramatically is that disabilities and their compensation are now more visible, but they are still not completely accepted. When Hockenberry appears on television giving his commentaries on the news, he is shown from the waist up. Perhaps the unremarked presence in a wheelchair of President Clinton's impressive counsel, Charles Ruff, at the impeachment trials of 1998-1999, will go far towards normalizing disabilities in the public eye.

For men with disabilities, the goal is to project masculine strength. But women with disabilities have also used overcoming adversity to enhance their self-image. Some women have reported feeling more capable and attractive in a wheelchair than on crutches (Lonsdale 1990, pp. 69-70). Nancy Mairs says she prefers to consider herself a cripple, rather than disabled or handicapped:

> People—crippled or not—wince at the word "cripple," as they do not at "handicapped" or "disabled." Perhaps I want them to wince. I want them to see me as a tough customer, one to whom the fates/gods/viruses have not been kind, but who can face the brutal truth of her existence squarely. As a cripple, I swagger (Mairs 1986, p. 9).

Mairs' presentation of self is "tough," a stance for women or men who want to confront the world on their own terms.

The 1996 Olympics in Atlanta became the scene of two dramas of socially transformed disabled bodies. The central moment of the opening ceremony is the lighting of the Olympic torch. This time the torch was lit by Muhammad Ali, the famous heavyweight champion and 1960 boxing gold medalist. Now he is weakened by Parkinson's disease—his left arm shook, his face was immobilized, and he could hardly walk. Why was he chosen to represent the spirit of athleticism when he seemed its very contradiction? As a man who was overcoming the limits of his body, he was celebrated once more as a hero (Vecsey 1996a).

A few days later, the Olympic audience saw that heroism in the face of physical trauma is not a masculine prerogative. Kerri Strug, an 18-year-old gymnast, took the final vault in the final team event on a badly sprained ankle, and clinched the first U.S. women's team gold medal. George Vecsey of *The New York Times* compared her to the mythic wounded soldier in war movies who leads the platoon to victory (Vecsey 1996b). The hero now was a woman. However, the contradictory image of her small body cradled in the arms of her burly coach was an apt symbolic representation of the gender contradictions of women with disabilities.

GENDER CONTRADICTIONS IN DISABILITY

The organized efforts of people with disabilities and their supporters have pressured governments to ensure that public spaces and transportation are accessible, that employers do not assume that they cannot handle disparate forms of work and that education be provided in ways that best address their needs. Both women and men have benefited from the successes of the disability rights movement. The special needs of women with disabilities for jobs, sexual relationships and a family life have not been so squarely faced.

In a widely cited paper, Michelle Fine and Adrienne Asch (1985) argued that women with disabilities face "sexism without the pedestal." Compared to men of a similar level of physical functioning, they are less likely to find jobs that allow them to be economically independent. They are also less likely to have a lifetime partner, because they need the care and attention that women are expected to give to others. Although there is little data on the effects of sexual orientation, they may be better able to find a life partner or a circle of women caregivers if they are lesbian.[3] Heterosexual women with disabilities, however, may find the fulfillment of traditional wife-mother gender roles beyond reach.

The opposite situation occurs for men with disabilities. They are more likely to find a life partner, a woman if they are heterosexual, a man if they are gay. In the traditional husband role, care received is reccompensed with economic support, so as long as a man with disabilities can earn an income, he can fulfill his family role obligations. Sexuality and its machismo qualities are particularly devastating minefields for men with disabilities, but it is also a problem for men who have had prostate surgery and older men. Indeed, the immense popularity of Viagra seems to indicate that actual or imagined sexual dysfunction is practically pandemic among men. In short, despite the conventional wisdom that " 'disabled man' is a self-contradiction, because men are stereotypically supposed to be 'able,' strong, and powerful," (Lakoff 1989, p. 68) a man with disabilities may be quite able to function well as a husband, father, and lover.

There is a loop-back effect between gender and disability. Family and work roles are gendered, and these affect the expectations for a person with disabilities. In turn, being a woman or a man with permanent disabilities modifies work opportunities, living arrangements, family life, friendships, intimate relationships and a person's sense of self (Charmaz 1995). The actual experience of women and men with disabilities often contradicts stereotypes of both gender and disability. However, the behavior of families, professional caretakers and physicians towards people with disabilities is more likely to reflect conventional ideas of what women and men should be. What the person with disabilities wants and can accomplish is encouraged or discouraged according to beliefs about "normal" gendered behavior.

The influence of gender expectations on caregiving and the socialization of children with a disability is clearly illustrated by a study of 32 low-income African American mothers of sons and daughters with sickle cell anemia (Hill and Zimmerman 1995). The children ranged in age from 2 months to 22 years, with an average age of 10.4 years. Through in-depth interviews, the researchers found that the mothers of daughters encouraged normal activities, including physical exertion, a stoical attitude toward minor symptoms, and self-care. In contrast, the mothers of sons described them "as fragile, certainly too fragile to conform to the traditional male gender role. The mothers tried to protect their sons from excessive physical activity, especially participation in sports" (p. 47). The mothers of sons with symptomatic sickle cell anemia were less likely to work outside the home than the mothers of daughters with the same level of functioning. They were more likely to do all they could to prevent their sons from suffering crises of pain and inability to breathe. In short, sickle cell anemia was not seen by these

African American mothers as preventing their daughters growing up be competent women, but was seen for their sons as "requiring behavi at odds with the aggressiveness, risk taking and physical activity of t male gender role" (p. 48). As a result, the mothers of daughters treate them the way they would any girl child, but the mothers of sons treate them as especially "vulnerable and in constant danger" (p. 48).

This study shows that caregivers start with an ideal of how wome and men should behave and then do an assessment of whether the pe son with disabilities can live up to that ideal. If the assessment is that t person cannot, then the means of living normally are denied. The ou come reinforces the gendered expectations rather than providing alte native ways of being a woman or a man. The content of the gend expectations is not the issue—beliefs that women can take care of ther selves or need protection, that men have to be physically assertive or a characterized by their earning ability vary by culture and social grou It is the power of the assumptions and their translation into behavior expectations for people with disabilities that create dual constraints f women and men who are especially dependent on others for person care and emotional support.

GENDERED CAREGIVING

The mainstreaming of people with disabilities depends on access technological devices and personal caregivers, who are often wome family members. Women have nursed sick family members as a lon standing part of their work as mothers, daughters and grandmothe (Wilkinson 1987). As professional and informal caregivers, women a the majority of the nurses and health-care workers in hospitals, nursir homes and in the home (Glazer 1990). Because it so closely resembl socially appropriate work, care is usually done by women for wome and men. Men with disabilities are more likely to be economically sel supporting and therefore more likely to have a wife to take care of the physical and emotional needs. Women with disabilities are less likely have a lifetime partner to look after them. They have to support them selves economically, and fend for themselves physically and emotio ally. They may even end up taking care of others.

A study of 25 middle-age women with a variety of disabilities (blin ness, hearing loss, polio, spinal cord injuries, cerebral palsy, rheum toid arthritis and multiple sclerosis), who needed help with dai activities, nevertheless "nurtured children, spouses, other family members, co-workers, and pets" (Quinn and Walsh 1995, p. 243). In a review of the autobiographies of 25 blind and visually impaired women and men, Adrienne Asch and Lawrence Sacks (1983) found that a common theme was that of the "self-sacrificing, supportive, nurturing mother." The men had one; the women became one. In adulthood, few of the women with visual impairments married, and most chose careers in teaching, social work and rehabilitation; the men attained high status in business or professions and married sighted women. Asch and Lawrence conclude that "the women provided support, warmth, and love for blind people just as their mothers did for them as blind children. The men ... tended to marry replicas of their mothers, whereas the women (symbolically) became their mothers" (Asch and Lawrence 1983, p. 244).

When women care for husbands with disabilities, they are doing what wives ordinarily do, only more extensively. A wife's attentiveness to a husband's needs, Kathy Charmaz points out, validates the centrality of his position in the household (1995). For men, caregiving is not the usual part of the husband or father role. As Robert Murphy, physically dependent on his wife because of a progressive spinal tumor, says: "Husbands become part-time nurses, which goes against social conventions, and wives find themselves with an additional child, which doesn't" (Murphy 1990, p. 206).

Yet the demands of caregiving are not gender-specific. Caregiving is a difficult combination of physical and emotional work—lifting, turning, toileting, feeding, bathing, clothing, encouraging, calming, hugging, kissing, talking (Corbin and Strauss 1988; Marks 1996).

Barbara Hillyer, reflecting on her care for her daughter, notes that "caring requires that exceptional physical and emotional strength be exercised. ... The ... caregiver is expected to be available, dependable and constant ..." well-organized, empathic but not expectant of emotional responsiveness from the recipient of care (Hillyer 1993, p. 11). It is the compatibility of caregiving duties with gendered family roles that makes it seem as if women are "natural" home nurses.

Women may be the designated caregivers, but they often find the burdens as difficult as men do. One study of spousal care gave quotes from eight wives and two husbands—the wives described excessive physical work, guilt feelings, hiding their worries, drinking too much, being constantly nervous, depressed or devastated; one husband talked of his resentment when he came home from work to a sick wife, the other of his sexual deprivation (Corbin and Strauss 1988, p. 289-317).

In one dramatic case, a caregiving husband, George Delury, age 62, helped his 52-year-old wife, Myrna Lebove, commit suicide with a drink of antidepressants, water and honey that he had mixed for her. They had been married after she developed her first symptoms of multiple sclerosis and had been together for 22 years. When her physical and mental condition deteriorated badly, Mr. Delury started a diary entitled "Countdown: A Daily Log of Myrna's Mental State and View Toward Death." In it, he said that he had four options—abandon his wife, keep taking care of her and go mad, kill himself or kill her. On the evidence of the diary, he was indicted for manslaughter, pleaded guilty, and was sentenced to six months in prison (Goldberg 1995; Pierre-Pierre 1996).

Although women are more likely to be family caregivers, there is ample evidence of men competently and willingly taking care of physically and mentally impaired relatives—women as well as men. A survey of a national sample of 233 mostly Caucasian men caregivers aged 36 to 84 showed that although they had not been socialized for the work, they were able to learn "on the job." The majority were caring for women with Alzheimer's disease. Like fathers who share parenting duties and single fathers of young children, men caregivers develop skills and empathy (Applegate and Kaye 1993).[4] They are, however, less likely to be competent in the routinized supervisory and concrete tasks that are so typical of "women's work:"

> In contrast to the stereotype that male caregiving is primarily instrumental in nature, these men's responses suggested that, overall, the tasks associated with social support were those they performed most frequently, most competently and with the greatest degree of satisfaction. Ranked second in degree of frequency, competence, and level of satisfaction were instrumental daily living tasks, followed by case management tasks and, in last place, the "hands-on" functional aspects of personal care (Kaye and Applegate 1990, p. 84).

The reluctance of the men in this study to handle a woman's body may be because many were caring for mothers. We would need more detailed accounts by men of their care of fathers, mothers, wives, and children of different ages and genders to have the full range of their caregiving activities to family members.[5] Outside of families, there is a great deal of evidence of the caregiving and emotional and social support of men to men ill with AIDS (Turner, Hays and Coates 1993).

GENDER AND ALTRUISM

When wives and mothers care for family members with disabilities, they are expanding already existing roles. But altruism seems to imbue women's overall behavior more than it does men's responses to calls to give. The gender effects evident in caregiving are highlighted in an extreme form of altruistic behavior—donation of a kidney to a relative with chronic renal disease. For the person who needs a kidney transplant, a successful donation means freedom from dialysis, and in many cases, survival. About 70 percent retain the kidney, and their subsequent physical and psychological quality of life is good (Simmons, Marine and Simmons 1987, pp. xxii-xxiii). The cost to the donor is loss of a healthy kidney with risk of subsequent failure of the remaining organ, surgery with general anesthesia, at least a week's hospitalization with consequent loss of stamina, and an extensive abdominal scar. The risk of the operation to a donor's life is calculated at .05 percent and the long-term risk at .07 percent (Simmons, Marine and Simmons 1987, pp. 39, 165-175).

A study of live kidney donors found that women are less likely to be ambivalent about donating a kidney to a relative with end-stage renal disease than men are (Simmons, Marine and Simmons 1987, pp. 188-189). Mothers asked to donate a kidney to a child are more likely to be free of doubt than fathers (58 percent to 29 percent), and sisters are more likely than brothers to be sure about their decision to donate (56 percent to 28 percent). Daughters agreeing to donate to a parent are surer that they are doing the right thing than are sons (27 percent to 11 percent).

After the kidney has been donated, men have more negative feelings than women do about what they have undergone, and these feelings persist a year later. However, men are more likely to feel better about themselves immediately post-transplant (23 percent to 8 percent) and one year later (40 percent to 26 percent), indicating that the women's donation may have been taken more for granted as part of their duty *as women*. The authors surmise that these gender differences are due to women's and men's experiences as actual or potential parents:

> Perhaps donation seems to the female to be a simple extension of her usual family obligations, while for the male it is an unusual type of gift. In our society the traditional female role is one in which altruism and sacrifice within the family is expected. ...

In fact, entering a hospital and suffering pain and body manipulation to give life to another is ordinarily a major part of her role and purpose in life. Giving birth to an infant is congruent psychologically with the act of giving a body part so a loved one can be reborn. ...

From a male's point of view, there is no life experience or expectation like childbirth that prepares him for this act of donation. Thus he may have stronger ambivalences and doubts even after the transplant. In any case, whether his feelings are positive or negative, he is more likely to feel he has performed an exceptional act. If his feelings are positive, as the majority of men's are, he is more likely to reap self-image benefits from this extraordinary gift (Simmons, Marine and Simmons 1987, pp. 188-189).

In this sense, women's bodily sacrifice is that of a "normal, natural" mother; men's image is similar to what is fostered when they are wounded in battle. They are heroes.[6]

A MAN BUT NOT A MAN, A WOMAN BUT NOT A WOMAN

Gendering of people with disabilities is especially evident in sexuality and procreation. Although many women with disabilities care for small children, the medical system and friends and family often discourage an active sex life and procreation for them, but encourage them for men (Gill 1994).[7] As a result, intimate partnerships between men with disabilities and able-bodied women are more prevalent than between women with disabilities and able-bodied men (Bullard and Knight 1981).

Conventional gender norms permeate sexual behavior, and social constructions of sexuality magnify the effects of physiological disabilities on sexual functioning.[8] Robert Murphy, a paraplegic, recognizes the variety of human sexual practices and pleasures, but he bleakly assesses the situation for paralyzed men:

> Most forms of paraplegia and quadriplegia cause male impotence and female inability to orgasm. But paralytic women need not be aroused or experience orgasmic pleasure to engage in genital sex, and many indulge regularly in intercourse and even bear children, although by Caesarean section. ... Paraplegic women claim to derive psychological gratification from the sex act itself, as well as from the stimulation of other parts of their bodies and the knowledge that they are still able to give pleasure to others. ... Males have far more circumscribed

anatomical limits. Other than having a surgical implant that produces a simulated erection, the man can no longer engage in genital sex. He either becomes celibate or practices oral sex—or any of the many other variations in sexual expression devised by our innovative species. Whatever the alternative, his standing as a man has been compromised far more than has been the woman's status. He has been effectively emasculated (Murphy 1990, p. 96).

In contrast, Irving Kenneth Zola, whose legs were wasted from polio, described mutual love-making with a severely paralyzed woman, an activity that began with his doing all her physical care, as well as the more conventional undressing, but nonetheless feeling that she had made love to him as well:

> And so the hours passed, ears, mouths, eyes, tongues inside one another. And every once in a while she would quiver in a way which seemed orgasmic. As I thrust my tongue as deep as I could in her ear, her head would begin to shake, her neck would stretch out and then her whole upper body would release with a sigh. ... So we ... hugged and curled up as closely as we could, with my head cradled in her arm and my leg draped across her. ... I fell quickly asleep ... rested, cared for, and loved (Zola 1982b, p. 216).

The difference in these two versions of sexuality lies in how love-making is defined—whether the physiological or emotional aspects are emphasized and whether sex is defined only as penile penetration. In Murphy's definition of male sexuality, a man who could not use his penis in intercourse is not really a man; Zola's definition is not only more egalitarian, its diffuseness encourages many forms of manhood—and womanhood. Those women and men with disabilities who are sexually experimental are more likely to have heterosexual and homosexual partners. Since women seem to be able to achieve orgasm through diffuse sexual excitement and men tend to focus their sexual behavior genitally, women with disabilities may have an advantage in that they are more willing to engage in a variety of sexual practices.

In other respects, they are disadvantaged. For women and men of any sexual orientation who have disabilities, social life is complicated not only by physiological differences that hamper conventional forms of sexual activity, but by difficulties in dating, going to parties and casual socializing. The problems of women with disabilities are frequently not physical but social—dating and being thought of in a sexual rather than platonic manner, especially during adolescence (Rousso 1988). They are likely to be stigmatized as potential sexual partners even when their disabilities are sensory—blindness and deafness (Becker and Jauregui 1985; Kolb 1985).

Women with disabilities tend to have the same sexual attitudes and desires as able-bodied women, but they are less likely to find heterosexual partners than men with disabilities (DeHaan and Wallander 1988). As for homosexual relationships, Rousso (1988) reports that lesbianism and bisexuality occurred among women with disabilities later in life and without a period of adolescent sexual exploration.

When they do have sexual relationships, women with disabilities find it difficult to obtain contraception or gynecological care commensurate with their physical needs, and their desire to have children is ignored or discounted (Killoran 1994; Waxman 1994). Sexual relations and fatherhood are considered normal aspirations for men with physical disabilities, but women with physical disabilities are often treated like asexual children. Carol Gill says that even after the passage of the American with Disabilities Act, "I am treated as though I don't belong with the other women who seek services in OB/GYN unless I can make my disability issues go away" (Gill 1994, p. 117).

Interviews with 31 women with a variety of disabilities, ranging in age from 22 to 69, well-educated and highly productive, and of different racial ethnic groups revealed "the common experience of having their reproductive needs undervalued. Many said their physicians treated them as asexual, thought they should not be having children and assumed that they would not be having children and that they did not want menstrual periods. In many cases, the first recommendation offered was to have a hysterectomy" (Nosek et al. 1995, p. 512).

Zola comments that "our society does not like to picture people who are weak, sick and even dying having needs for sexual intimacy" (Zola 1982a, p. 214). He found that there were few apartments for couples in an otherwise comprehensive living facility in the Netherlands, and that a Swedish project that encouraged the sexual involvement of able-bodied counselors with people with severe disabilities was discontinued because the counselors "...actually began to find these very physically disabled people attractive, and *that* was regarded as shocking if not sick."

A recognition of the sexuality of women and men with long-term physiological problems would go a long way to eroding the stereotypes of both gender and disability. Such rethinking involves breaking down the conventional categorization of women with disabilities as asexual and childlike and men with disabilities as severely frustrated in their sexual expression. In a more general sense, the same depolarization of men versus women and disabled versus able-bodied is necessary to change the social status of those with long-term physiological problems from "outsider" to "one of us."

TOWARDS A CONTINUUM OF BODIEDNESS

One feminist view of women with disabilities argues that they are "doubly handicapped"—by their gender and by their physical limitations (Deegan and Brooks 1985, Morris 1993). In this view, a physical disability is a "minority" status added to all the other jeopardies suffered (Deegan 1985). Thus, an African American man with a disability would also be doubly disadvantaged.

Minority-group status allows for activism and fighting for civil rights on the basis of shared interests that enlist people with very different needs. The disability rights movement in the United States successfully fought for ramps in all public buildings, wheelchair-accessible bathrooms, wheelchair lifts in buses, apartments equipped for independent living, braille numbers and bells to indicate floor stops in elevators, closed-captioned television programs, infrared hearing devices in theaters, sign-language translation at public forums, telephone devices and communications relays and so on. The movement has also produced anti-discrimination legislation for jobs and housing to integrate people with disabilities into mainstream life. However, just as feminism has been accused of not attending to the multiple disadvantages of women of color and lesbians, a broad-based disability rights movement does not address gender differences any more than it does racial or sexual differences that compound discriminatory treatment (Wendell 1992).

Whether physically challenged women or men can work at paid jobs and care for themselves and others at home depends to a great extent on the availability of technological devices, transportation, the physical environment at work, in houses and apartments and in stores, but perhaps more on the active participation of employers, families, and friends (Fifield et al. 1989; Russo and Jansen 1988). The goal is to find alternative ways to accomplish the tasks of daily living, ways that may be different for women and men when their family situations are different. When their goals are the same—to hold down a paying job, for example—women with disabilities may need more of what women in general need to counter both sexism and prejudice against those with physical impairments. A determined effort at inclusion of people of different genders, racial ethnic identities, parental status and sexual orientation, as well as different ranges of bodily capabilities, is more effective here than a stance of neutrality. These efforts must include changing the workplace environment on all counts—scheduling, work assignments and responses of supervisors, subordinates and co-workers—to ensure that the inclusion is more than a superficial, showcase phenomenon.

Just as the integration of people with disabilities into the mainstream of society depends on attention to their bodily needs by environmental modifications, access to technology and and personal-care attendants, the special biological needs of women with disabilities cannot be ignored. Menstruation, contraception, pregnancy and childbirth are not problems to be done away with by early hysterectomies, but challenges to health professionals. The extensive repertoire of technologies to assist procreation in women (and men) with infertility problems should be available to those with other disabilities as well. Despite evidence that women with disabilities can run households, take care of children and work outside the home, the perception persists that such tasks are impossible for her (Killoran 1994; Shaul, Dowling and Laden 1985). Women with quite severe physical limitations have devised ways of caring for small children: "One mother who could not use her arms found that her two children both learned to scramble up her and hang around her neck" (Lonsdale 1990, p. 79). Carrie Killoran says of her own parenting, "People with disabilities are already accustomed to doing everything differently and more slowly, and caring for children is no different" (Killoran 1994, p. 122; Finger 1990; Kocher 1994).

The conventional norms of femininity have locked women with disabilities into a paradoxical situation—as women, it is all right for them to be helpless and dependent, but, because they are disabled, they are unlikely to have a man to take care of them. Feminists have argued that norms of independence and economic self-support provide a better model for all women, and that giving women with disabilities the means to accomplish these goals would go a long way to enhancing their self-esteem and quality of life (Asch and Fine 1988).

For men with disabilities, the change has to come in challenges to conventional masculinity. Men could expand their options by having different kinds of relationships rather than overburdening one caregiving woman. Both women and men might welcome group living, but it would have to be autonomous and without restrictions on sexual coupling and children.

Gender expectations and assumptions are harder to change than the physical environment and job requirements. If social groups parcel out roles on the basis of gender, then identity as a woman and a man are tied to being able to fulfill these gender-appropriate roles. Looking at the problem of masculinity and physical disability in the lives of 10 men, Thomas Gerschick and Adam Stephen Miller (1994) discovered three strategies: reliance on conventional norms and expectations of manhood, reformulation of these norms and creation of new norms. The men who relied on the predominant ideals of masculinity felt they had to demonstrate physical strength, athleticism, sexual prowess and independence. Their self-image was tied to heroics and risk-taking, but they often felt inadequate and incomplete because they couldn't do what they wanted or go where they wanted. The men who reformulated these norms defined their ways of coping with their physical limitations as demonstrations of strength and independence. For example, two quadriplegics who needed round-the-clock personal care assistants did not feel they were dependent on others, but had hired helpers whom they directed and controlled. The men who rejected the standard version of masculinity put more emphasis on relationships than on individual accomplishments, were comfortable with varieties of sexuality, and felt they were non-conformists.

To erase the status dilemmas of women and men with physical disabilities, conventional norms about bodies, functions, beauty and sexuality need to be re-examined (Asch and Fine 1988; Wendell 1996). Making the experiences of women and people of color visible forced a reconsideration of stereotypes of normality and otherness; similarly, because "physically disabled people have experiences which are not available to the able-bodied, they are in a better position to transcend cultural mythologies about the body" (Wendell 1992, p. 77). Few people are as beautiful as movie stars or as muscular as body builders. A woman without arms or legs claimed the statue of Venus de Milo as her model of beauty (Frank 1988). Orgasms can be felt in many parts of the body other than the genitals. Races can be run in wheelchairs as well as on foot, on horses, on bicycles and in cars.

Bodies matter to the person with pain, limited mobility and sensory difficulties, but the way they matter is also a social phenomenon (Butler

1993; Wendell 1996). One answer to the status contradictions imposed on people with disabilities is to discard the concept of "other." A study of nondisabled people who had long-term relationships with people with extremely severe disabilities found that the "partnership" was based on a sense of essential humanity and full integration into each other's social space (Bogdan and Taylor 1989). Ability and disability, bodily integrity and bodily dysfunction and standards of beauty are all relative. The variety of bodies and social environments make all of us part of a complex continuum of able-bodiness, just as the variety of women and men calls into question gender stereotypes.

ACKNOWLEGMENTS

I am indebted to Susan A. Farrell and Barbara Katz Rothman for help in revising this paper.

NOTES

1. An Israeli study found that parents were more likely to reject a newborn with what they felt were "non-human" facial or bodily deformities that were non-life-threatening than one with serious, but hidden, defects (Weiss 1994).
2. Their social acceptance is only on the surface: "As with the poor relation at the wedding party, so the reception given the handicapped person in many social situations: sufficient that he is here, he should not expect to dance with the bride" (Davis 1972, p. 140).
3. For a beautiful account of caregiving by the lesbian partner of a woman dying of cancer and the circle of women friends they gathered to help them, see Butler and Rosenblum (1991).
4. On men as intimate parents, see Risman (1987).
5. For a poignant, detailed account by a man of his care of his wife in the last stages of Alzheimer's, see Bayley (1998).
6. I am indebted to Susan Farrell for this point.
7. See "Women with Disabilities: Reproduction and Motherhood," edited by Marsha Saxton. *Sexuality and Disability,* summer 1994.
8. For personal testimonies and counseling advice on these issues, see Bullard and Knight (1981).

REFERENCES

Applegate, J.S., and L.W. Kaye. 1993. "Male Elder Caregivers." In *Doing "Women's Work:" Men in Non-traditional Occupations,* edited by C.L. Williams. Thousand Oaks, CA: Sage.

Asch, A., and M. Fine. 1988. "Introduction: Beyond Pedestals." In *Women with Disabilities: Essays in Psychology, Culture, and Politics*, edited by M. Fine and A. Asch. Philadelphia, PA: Temple University Press.
Asch, A., and L.H. Sacks. 1983. "Lives Without, Lives Within: Autobiographies of Blind Women and Men." *Journal of Visual Impairment and Blindness* 77: 42-47.
Bayley, J. 1998. *Elegy for Iris*. New York: St. Martin's Press.
Becker, G., and J.K. Jauregui. 1985. "The Invisible Isolation of Deaf Women: Its Effect on Social Awareness." In *Women and Disability: The Double Handicap*, edited by M.J. Deegan and N.A. Brooks. New Brunswick, NJ: Transaction Books.
Bogdan, R., and S.J. Taylor. 1989. "Relationships with Severely Disabled People: The Social Construction of Humanness." *Social Problems* 36: 135-48.
Bullard, D.G., and S.E. Knight. 1981. *Sexuality and Physical Disability: Personal Perspectives*. St. Louis, MO: Mosby.
Butler, J. 1993. *Bodies that Matter: On the Discursive Limits of "Sex."* New York and London: Routledge.
Butler, S., and B. Rosenblum. 1991. *Cancer in Two Voices*. San Francisco: Spinster Book Company.
Charmaz, K. 1995. "Identity Dilemmas of Chronically Ill Men." In *Men's Health and Illness: Gender, Power and the Body*, edited by D. Sabo and D.F. Gordon. Thousand Oaks, CA: Sage.
Corbin, J.M., and A. Strauss. 1988. *Unending Work and Care: Managing Chronic Illness at Home*. San Francisco, CA: Jossey-Bass.
Davis, F. 1972. "Deviance Disavowal: The Management of Strained Interaction by the Visibly Handicapped." In *Illness, Interaction and the Self*. Belmont, CA: Wadsworth.
Deegan, M.J. 1985. "Multiple Minority Groups: A Case Study of Physically Disabled Women." In *Women and Disability: The Double Handicap*, edited by M.J. Deegan and N.A. Brooks. New Brunswick, NJ: Transaction Books.
Deegan, M.J., and N.A. Brooks. 1985. "Introduction—Women and Disability: The Double Handicap." In *Women and Disability: The Double Handicap*, edited by M.J. Deegan and N. A. Brooks. New Brunswick, NJ: Transaction Books.
DeHaan, C.B., and J.L. Wallander. 1988. "Self-concept, Sexual Knowledge and Attitudes, and Parental Support in the Sexual Adjustment of Women with Early- and Late-onset Physical Disability." *Archives of Sexual Behavior* 17: 145-161.
Fifield, J., S. Reisine, C.A. Pfeiffer, and G. Affleck. 1989. "Workplace Disability: Gender, Technology, and the Experience of Rheumatoid Arthritis." In *Healing Technology: Feminist Perspectives*, edited by K.S. Ratcliff. Ann Arbor, MI: University of Michigan Press.
Fine, M., and A. Asch. 1985. "Disabled Women: Sexism Without the Pedestal." In *Women and Disability: The Double Handicap*, edited by M.J. Deegan and N.A. Brooks. New Brunswick, NJ: Transaction Books.
Finger, A. 1990. *Past Due: A Story of Disability, Pregnancy and Birth*. Seattle, WA: Seal Press.
Frank, G. 1988. "On Embodiment: A Case Study of Congenital Limb Deficiency in American Culture." In *Women with Disabilities: Essays in Psychology, Culture, and Politics*, edited by M. Fine and A. Asch. Philadelphia, PA: Temple University Press.

Gallagher, H.G. 1985. *FDR's Splendid Deception.* New York: Dodd Mead.

Gerschick, T.J., and A.S. Miller. 1994. "Gender Identities at the Crossroads of Masculinity and Physical Disability." *Masculinities* 2: 34-55.

Gill, C.J. 1994. "Editorial: When is a Woman Not a Woman." *Sexuality and Disability* 12: 117-119.

Glazer, N. 1990. "The Home as Workshop: Women as Amateur Nurses and Medical Care Providers." *Gender & Society* 4: 479-499.

Goldberg, C. 1995. "Suicide's Husband is Indicted: Diary Records Pain of 2 Lives. *The New York Times* (Dec. 15).

Hill, S.A., and M.K. Zimmerman. 1995. "Valiant Girls and Vulnerable Boys: The Impact of Gender and Race on Mothers' Caregiving for Chronically Ill Children." *Journal of Marriage and the Family* 57: 43-53.

Hillyer, B. 1993. *Feminism and Disability.* Norman, OK: University of Oklahoma Press.

Hockenberry, J. 1995. *Moving Violations: War Zones, Wheelchairs, and Declarations of Independence.* New York: Hyperion.

Hughes, E.C. 1971. "Dilemmas and Contradictions of Status." In *The Sociological Eye.* Chicago and New York: Aldine Atherton.

Kaye, L.W., and J.S. Applegate. 1990. *Men as Caregivers to the Elderly: Understanding and Aiding Unrecognized Family Support.* Lexington, MA: Lexington Books.

Kelly, M. 1996. "Accentuate the Negative." *The New Yorker* (April 1): 44-48.

Killoran, C. 1994. "Women with Disabilities Having Children: It's Our Right Too." *Sexuality and Disability* 12: 121-126.

Kocher, M. 1994. "Mothers with Disabilities." *Sexuality and Disability* 12: 127-133.

Kolb, C. 1985. "Assertive Training for Women with Visual Impairments." In *Women and Disability: The Double Handicap,* edited by M.J. Deegan and N. A. Brooks. New Brunswick, NJ: Transaction Books.

Lakoff, R.T. 1989. "Review Essay: Woman and Disability." *Feminist Studies* 15: 365-375.

Lonsdale, S. 1990. *Women and Disability.* New York: St. Martin's Press.

Mairs, N. 1986. *Plaintext.* Tucson: University of Arizona Press.

Marks, N.F. 1996. "Caregiving Across the Lifespan: National Prevalence and Predictors." *Family Relations* 45: 27-36.

Morris, J. 1993. "Feminism and Disability." *Feminist Review* 43: 57-70.

Murphy, R.F. 1990. *The Body Silent.* New York: Norton.

Nosek, M.A., M.E. Young, D.H. Rintala et al. 1995. "Barriers to Reproductive Health Maintenance Among Women with Physical Disabilities." *Journal of Women's Health* 4: 505-518.

Pierre-Pierre, G. 1996. "Man Who Helped Wife Die to Serve 6 Months." *The New York Times* (May 18).

Quinn, P., and S.K. Walsh. 1995. "Midlife Women with Disabilities: Another Challenge for Social Workers." *Affilia* 10: 235-254.

Risman, B.J. 1987. "Intimate Relationships from a Microstructural Perspective: Men who Mother." *Gender & Society* 1: 6-32.

Rousso, H. 1988. "Daughters with Disabilities: Defective Women or Minority Women? In *Women with Disabilities: Essays in Psychology, Culture, and Politics*, edited by M. Fine and A. Asch Philadelphia PA: Temple University Press.
Russo, N.F., and M.A. Jansen. 1988. "Women, Work, and Disability: Opportunities and Challenges." In *Women with Disabilities: Essays in Psychology, Culture, and Politics*, edited by M. Fine and A. Asch. Philadelphia, PA: Temple University Press.
Shaul, S., P.J. Dowling, and B.F. Laden. 1985. "Like Other Women: Perspectives of Mothers with Physical Disabilities." In *Women and Disability: The Double Handicap*, edited by M.J. Deegan and N.A. Brooks. New Brunswick, NJ: Transaction Books.
Simmons, R.G., S.K. Marine, and R.L. Simmons. 1987. *Gift of Life: The Effect of Organ Transplantation on Individual, Family, and Societal Dynamics*. New Brunswick, NJ: Transaction Books.
Thomson, R.G. 1994. Review essay: "Redrawing the Boundaries of Feminist Disability Studies." *Feminist Studies* 20: 583-595.
Turner, H.A., R.B. Hays, and T.J. Coates. 1993. "Determinants of Social Support Among Gay Men: The Context of AIDS." *Social Problems* 34: 37-53.
Vecsey, G. 1996a. "Choosing Ali Elevated These Games." *The New York Times* (July 21).
_____. 1996b. "Strug Took Her Chances for the Gold." *The New York Times* (July 24).
Waxman, B.F. 1994. "Up Against Eugenics: Disabled Women's Challenge to Receive Reproductive Services." *Sexuality and Disability*, 12: 155-171.
Weiss, M. 1994. *Conditional Love: Parental Relations Toward Handicapped Children*. Westport, CT: Greenwood Press.
Wendell, S. 1992. "Toward a Feminist Theory of Disability." In *Feminist Perspectives in Medical Ethics*, edited by H.B. Holmes and L.M. Purdy. Bloomington, IN: University Press.
_____. 1996. *The Rejected Body: Feminist Philosophical Reflections on Disability*. New York and London: Routledge.
Wilkinson, D.Y., and M.B. Sussman. 1987. *Alternative Health Maintenance and Healing Systems for Families*. New York: Haworth.
Zola, I.K. 1982a. *Missing Pieces: A Chronicle of Living with a Disability*. Philadelphia, PA: Temple University Press.
_____. 1982b. "Tell Me, Tell Me." In *Ordinary Lives: Voices of Disability and Disease*, edited by I.K. Zola. Cambridge, MA: Apple-wood Books.

SUCCESSFUL LABOR MARKET TRANSITIONS FOR PERSONS WITH DISABILITIES
FACTORS AFFECTING THE PROBABILITY OF ENTERING AND MAINTAINING EMPLOYMENT

Edward Yelin and Laura Trupin

ABSTRACT

This paper uses data from the Annual March Supplement to the Current Population Survey (CPS) to provide contemporary estimates of employment rates among persons with disabilities, to estimate the fraction of such persons who did not work in the year prior to the survey but were able to enter jobs, to estimate the fraction of those who worked in the year prior to survey but who left jobs and to analyze the factors affecting the probability that persons with disabilities will be able to enter new jobs or maintain the ones they hold.

We find that persons with disabilities are about 30 percent as likely to be employed at any one time as persons without disabilities, if unemployed they are about one-fifth as likely to enter jobs and if employed they are about three times as likely to leave work. Differences between persons with and without disabilities in demographic and work characteristics account for a substantial fraction of the gap in their employment rates; a significant, albeit smaller, fraction of the difference in their ability to maintain jobs they already hold; and almost none of the difference in their ability to gain entry to new jobs. Disability, thus, would appear to account for low rates of job entry among persons with disabilities, but low employment rates and high rates of job loss among such persons are apparently due in large measure to other demographic and occupational factors.

INTRODUCTION

Employment rates among persons with disabilities are low relative to the rates among persons without disabilities (Yelin 1996). Therefore, to improve the employment picture for persons with disabilities it is necessary to increase the proportion of such persons who enter new jobs, while reducing the proportion who leave work. In this paper, we focus on the characteristics of persons with disabilities and of labor markets which are associated with increased rates of job entry and reduced rates of job loss. In doing so, we attempt to estimate the extent to which these rates are associated with disability per se, as opposed to the other characteristics of persons with disabilities that may account for their employment problems: for example, gender, race, and educational status (Trupin et al. 1997).

The proportion of persons with disabilities in the labor force has shifted substantially over rather short periods of time, apparently in concert with major trends in labor force participation, rather than with the prevalence of potentially disabling conditions (Yelin 1992; Stapleton et al. 1994). These major trends include a decrease in labor force participation among men, particularly older and nonwhite men, and an increase among women, particularly younger and white women (Yelin and Katz 1994a). The fate of persons with disabilities is also tied to the transformation of the economy from one based on the production and distribution of manufactured goods to one based on services. Women with disabilities have benefited from the expansion of the service sector, while men with disabilities have experienced a disproportionate share of the decline in manufacturing jobs (Yelin 1992). Similarly, persons with disabilities have sustained a disproportionate share of the growth in part-time employment, particularly involuntary part-time employment (Yelin and Katz 1994b). The analyses of the impact of the labor market on persons with disabilities are consistent with the notion that disability—like race, gender and age—places an individual at the end of the labor market queue, able to enter when opportunities abound, forced to leave when the demand for labor slackens. Indeed, there is evidence that disability interacts with race, gender and age to reduce the labor force participation rate more than would be expected on the basis of these characteristics alone (Trupin et al. 1997).

In the initial set of analyses, we show the employment rate of persons with and without disabilities, as well as the frequency with which each group enters new jobs or maintains old ones. Subsequently, we use the results of multivariate logistic regressions to indicate the extent to which differences in the employment rates and transitions into and out of employment by persons with and without disabilities is due to disability status or demographic characteristics or occupational factors. Finally, we use multivariate logistic regressions to estimate the impact of demographic and occupational factors on the probability of transitions into and out of employment; these latter regressions are estimated separately for persons with and without disabilities to highlight differences in the two groups in the factors affecting labor market success.

METHODS

The data source for the analyses reported here is the annual March Supplement to the monthly Current Population Survey (CPS) for the years 1993 to 1996. The CPS is the principal venue for the estimation of the monthly national employment statistics; the annual March Supplement provides information on labor force participation in the entire year prior to survey including occupation, industry, weeks of work per year and hours of work per week. In addition, the supplement includes extensive information on the demographic characteristics of the respondents and their families, the geographic areas in which they reside and the earnings and income of each family member by source (U.S. Bureau of the Census 1993).

In each year, the CPS contains information about approximately 57,000 households with about 112,000 individuals age 15 and older, as well as 33,500 children. The analyses reported here are limited to the approximately 93,000 individuals of working ages—18 through 64—in each year of the CPS.

In the CPS, disability is defined by the answers to questions that ask whether respondents have health limitations that prevent work or limit the amount or kind of work. This is a fairly severe definition of disability, and relatively small proportions of the respondents (fewer than 8 percent) meet the definition in any one year. To increase the statistical power of the descriptive and analytic estimations among the subsample with disabilities, we have therefore merged the four years of the CPS.

Analyses

In the initial set of analyses, we provide estimates for 1993 through 1996 of the number and percentage of persons with and without disabilities who were not employed during the entire year prior to survey, but who were working when interviewed (a measure of those entering jobs); of the number and percentage of such persons who were employed at some point during the year prior to survey but were not working when interviewed (a measure of those leaving jobs); and of the number and percentage employed regardless of prior year status. These estimations are based on bivariate cross-tabulations. In order to determine the extent to which the employment situation of persons with disabilities is due to the disability or to other characteristics, we then use logistic regression to calculate an adjusted probability of entering jobs, leaving jobs and being employed after taking into account disability status, demographic and regional characteristics of the CPS respondents and, where appropriate, the characteristics of their work in the prior year.

In the final set of analyses, we use logistic regression to estimate the factors affecting the ability of persons with disabilities to enter and maintain employment. A separate set of estimations was conducted on a data partition including only persons without disabilities.

In the multivariate regression models, the demographic characteristics include age (indicator variables for 18-24, 25-34, 35-44, 45-54, with 55-64 as the referent), gender, white versus nonwhite race, Hispanic versus non-Hispanic status, marital status (indicator variables for being currently married and widowed, separated or divorced, with never married as the referent), type of household (living in a household with a male head or female head, with both a male and female adult present as the referent), education (indicator variables for less than high school, high school graduate, some college or college graduate, with graduate school or more as the referent), Census region of the country (indicator variables for Midwest, South or West, with Northeast as referent) and type of residential environment (indicator variables for suburb, rural area or small city, with central city as the referent).

In the regression models estimating the probability of maintaining jobs the following work characteristics were included: occupation (indicator variables for technical or sales, administrative, service, crafts, operatives, transportation and laborers, with executive/professional occupations as the referent); industries (indicator variables for agriculture-mining-construction, manufacturing, transportation, communication, utilities, wholesale and retail trades, finance, insurance and real estate, services and government, with professional services as the referent); size of firm (indicator variables for fewer than 10 workers, 11-24, 25-99, 100-499 and 500-999, with greater than 1,000 as the referent); hours of work in the typical week; whether the employer offers health insurance; whether the employer or union offers and/or the employee utilizes a pension plan; the personal earnings of the respondent (in increments of $10,000); the personal earnings of the remainder of the household's members (by $10,000s); and the total of all forms of income other than earnings within the household (also in $10,000s).

In the regression models estimating the probability of entering jobs, three income variables were included: the personal earnings of all household members (in $10,000s), disability payments of all household members (in $10,000s), and the total of all other forms of household income (in $10,000s).

Each of the regressions also included indicator variables for the year in which the data were collected, with 1993 serving as the referent.

Limitations

Many labor market analysts use the Current Population Survey because of its geographic coverage and because the sample size is sufficiently large to allow statistically robust estimations among relatively small population subgroups, especially when multiple years are merged. However, the basic design of the CPS limits the kinds of labor market analyses that may be performed and, more importantly for this paper, the definition of disability is inconsistent with contemporary notions.

As to the former point, when individuals are interviewed for the March Supplement, they report whether they worked at any point in the year prior to survey and whether they worked in the past two weeks. Since the probability of employment in an entire year is much higher than the probability in any two-week period, the number of persons estimated to leave work will appear higher than the number entering jobs as an artifact of the sampling design.

There is also no information on any employment that precedes the year prior to the survey other than whether the individual last worked for pay, if ever. Thus, for those who did not work at all during the year prior to the survey, there is no indication of the extent of employment experience or previous occupations or industries, all of which may affect the likelihood of entering the labor market.

In addition, no information is available in the CPS on the two-month period between the end of the prior year and the time of interview for the March Supplement. Because of this gap, the analyst is forced to infer transitions into and out of employment. Finally, in 1994, the CPS changed the questions used to establish labor force participation rates (Polivka and Rothgeb 1993), rendering precise comparisons of employment status prior to and after that date impossible.

As to the latter point, the Americans with Disabilities Act of 1990, reflecting the notion that disability results when there is a mismatch between the individual's functional capacity and environmental accommodations, mandates reasonable accommodation to physical and mental impairment (Jones 1991). In the CPS, respondents merely report whether they have health limitations which prevent or limit the amount or kind of work; they are not asked to report work capacity in the presence or absence of accommodations. Accordingly, the estimate of the prevalence of disability and the estimate of employment among persons with disabilities do not indicate work capacity under the assumption of strict enforcement of the employment provisions of the ADA.

Results

Table 1 presents the tabulations of the number and percentage of persons with and without disabilities who entered jobs, left jobs and were currently employed in the years 1993 to 1996. The first set of results

Table 1. No. (in Millions) and Percent of Persons Entering Jobs, Leaving Jobs, and Currently Employed, by Disability Status and Year, U.S., 1993-1996

	Job Entrants Among Persons Not Working Prior Year						Job Leavers Among Persons Working Prior Year						Currently Employed as of Date of Survey					
	Persons w. Disabilities		Persons w.o. Disabilities				Persons w. Disabilities		Persons w.o. Disabilities				Persons w. Disabilities		Persons w.o. Disabilities			
Year	N	%	N	%		Ratio	N	%	N	%		Ratio	N	%	N	%		Ratio
1993	0.113	1.5	1.621	6.9		.22	1.965	45.3	23.345	19.3		2.35	2.491	21.1	99.155	68.6		.31
1994	0.117	2.0	2.626	11.0		.18	1.869	42.8	17.240	14.1		3.04	2.674	21.0	107.939	73.7		.29
1995	0.157	1.9	2.312	10.1		.19	1.918	42.2	16.468	13.2		3.20	2.784	21.7	110.388	74.8		.29
1996	0.170	2.0	2.289	10.0		.20	1.731	39.8	16.578	13.2		3.02	2.792	22.0	111.549	75.0		.29

Source: Authors' analysis of the March Supplement to the Current Population Survey, 1993-1996.

indicate that the number of persons with disabilities who entered jobs that is the number who had not worked in the year prior to survey and were working as of the interview for the March Supplement to the CPS is small—varying between 117,000 and 170,000 a year after the changes in the CPS were implemented in 1994. This represents approximately 2 percent of all persons with disabilities who had not worked in the year prior to survey. Persons with disabilities were only about one fifth as likely to enter jobs in any one year as those without disabilities.

The second set of results in Table 1 indicates that about 40 percent of persons with disabilities who had worked in the prior year had left work by the time of the March Supplement interviews for 1994 to 1996 (the percentage was slightly higher in 1993, probably reflecting both the labor market questions used in the CPS through that year and the economic climate in the nation). Persons with disabilities were more than three times as likely to leave jobs as those without disabilities during the period 1994 to 1996. All told, in excess of 1.7 million persons with disabilities in each March Supplement reported having worked in the year prior to survey and being unemployed as of the interview date.

Net of the flows into and out of employment, at any one time only slightly more than one-fifth of persons with disabilities of working age reported being employed in each year (third set of results in Table 1). In 1996, this represented about 2.8 million persons. In contrast, three quarters of persons without disabilities were employed as of that year. Thus, persons with disabilities are only about 30 percent as likely to be employed as those without, and the labor market is numerically dominated by persons without disabilities, of whom in excess of 111 million were working in 1996.

There is an extensive literature documenting factors other than disability which affect the probability of labor force participation. Table 2 presents estimates of the magnitude of the gap between persons with and without disabilities in the proportion who enter and leave jobs and are employed at any one time that remains after statistical adjustment for many of these factors. Recall from Table 1 that, on an unadjusted basis, persons with disabilities were about 20 percent as likely as those without disabilities to enter jobs in any one year. Statistical adjustment had little effect on the proportion of persons with and without disabilities entering jobs (first set of columns in Table 2), suggesting that it is disability itself rather than the demographic characteristics of persons

Table 2. Percent of Persons Entering Jobs, Leaving Jobs, and Currently Employed, after Adjustment for Demographic and Work Characteristics[a], by Disability Status and Year, U.S., 1993-1996

	Job Entrants Among Persons Not Working Prior Year			Job Leavers Among Persons Working Prior Year			Currently Employed as of Date of Survey					
	Adjusted for Demographic Characteristics			Adjusted for Demographic and Work Characteristics			Adjusted for Demographic Characteristics			Adjusted for Demographic Characteristics, and Prior Year Employment Status		
	Persons w. Disabilities	Persons w.o. Disabilities		Persons w. Disabilities	Persons w.o. Disabilities		Persons w. Disabilities	Persons w.o. Disabilities		Persons w. Disabilities	Persons w.o. Disabilities	
Year	%	%	Ratio	%	%	Ratio	%	%	Ratio	%	%	Ratio
1993	1.4	6.7	.21	40.4	19.0	2.13	21.3	68.3	.31	40.3	66.4	.61
1994	2.3	10.3	.22	32.7	14.2	2.30	25.4	73.0	.35	47.9	71.1	.67
1995	2.1	9.5	.22	31.7	13.6	2.33	26.4	73.9	.36	48.6	71.6	.68
1996	2.1	9.4	.22	31.9	13.7	2.33	26.5	74.0	.36	48.7	71.6	.68

Notes: [a]Demographic variables in models include age, gender, race, ethnicity, marital status, household type, education, region of country, and residential environment. Work variables include occupation, industry, size of firm, work hours, employee health and retirement benefits, individual and household earnings, and household nonearned income.

Source: Authors' analysis of the March Supplement to the Current Population Survey, 1993-1996.

with disabilities that accounts for most of the difference in the rate of job entry.

On the other hand, statistical adjustment for demographic and work characteristics substantially reduced differences between persons with and without disabilities in the proportion of each group who had worked in the year prior to survey and were not working as of the date of the March Supplement interview (second set of columns in Table 2). Thus, on an unadjusted basis, persons with disabilities were more than three times as likely to leave jobs as those without disabilities from 1994 to 1996 (Table 1). After adjustment, they were about 2.3 times as likely to stop working in those years (adjustment also reduced the gap in 1993, prior to the change in the CPS labor force participation items).

In the third and fourth set of columns in Table 2, we present estimates of the proportion of persons with and without disabilities who reported being employed as of the March Supplement interviews from 1993 through 1996, adjusted for demographic characteristics alone (third set of columns) and for demographic characteristics and prior-year employment status. Adjustment for demographic characteristics alone reduced the difference in employment rates between persons with and without disabilities substantially. Recall from Table 1 that persons with disabilities were about 22 percent as likely to be employed as persons without disabilities for each of the years 1994 through 1996; after adjustment for demographic characteristics, persons with disabilities were between 35 percent and 36 percent as likely to be employed. However, adjustment for prior-year work status in addition to demographic characteristics had a stronger effect than demographic characteristics alone, increasing the ratio of employment rates among persons with and without disabilities from about 35 percent (Table 1) to between 67 percent and 68 percent (Table 2, fourth set of columns) for the period 1994 through 1996.

Thus, part of the gap in employment in any one year is due to demographic characteristics, but far more is due to prior-year work status. The findings about the importance of demographic characteristics and prior-year work status are consistent with those in the literature concerning employment among persons with disabilities associated with specific conditions (Yelin et al. 1987; Reisine et al. 1989; Murphy 1991; Blanc et al. 1993) and with a wide range of conditions (Yelin and Katz 1994a). That prior-year work status affects current-year employment status so strongly suggests the importance of helping persons with disabilities gain initial entry to employment; the low rate of job entry, even after adjustment for demographic characteristics, testifies to the difficulty of achieving this goal.

There is an extensive literature documenting the strong association between low social class, whether measured by income, occupation or education, and poor health status (Wilkinson 1996; Adler and Matthews 1994). The relationship between social class and disability status has also been well documented (Berkowitz et al. 1976; Levitan and Taggart 1977; LaPlante and Carlson 1996). Finally, researchers have also established that persons with such characteristics as low levels of education—regardless of disability status—are less likely to succeed in the labor force than those from higher social classes (Jencks et al. 1988), and that persons with disabilities are more likely to have the characteristics that reduce labor force participation (Yelin 1996). It is not surprising, therefore, that differences between persons with and without disabilities in demographic and work characteristics should account for part of the difference in employment between the two groups.

PROBABILITY OF ENTERING AND LEAVING JOBS

In Tables 3 through 8, we present the results of the multivariate regressions estimating the impact of demographic and work characteristics on the probability of entering or maintaining jobs, two definitions of success in the labor market. In each of the tables, we present the probability of entering (or maintaining) jobs for each category of selected demographic or work characteristic after adjustment for the remaining characteristics in the model. In addition, we show the ratio of the probability of entering (or maintaining) jobs across categories of each characteristic; this is a measure of relative risk. For those characteristics for which there are more than two categories, such as education levels, we show the relative risk for the lowest and highest levels.

Table 3 highlights differences by demographic characteristics in the probability of job entry *among persons with disabilities*. Differences in the probability of job entrance between men and women and between Hispanics and non-Hispanics with disabilities did not reach traditional levels of statistical significance, although men were slightly more likely than women and non-Hispanics slightly more likely than Hispanics to enter jobs. Surprisingly, the probability of job entrance did not differ significantly from the reference education category—graduate educa-

Table 3. Probability of Job Entrance for Persons with Disabilities, by Selected Demographic Characteristics, with Adjustment for Demographic Characteristics[a], U.S. Average for 1993-1996

Demographic Characteristic	Persons with Disabilities
Gender	
Male	2.0%[NS]
Female	1.8%
Ratio	1.11
Race	
White	2.0%
Nonwhite	1.4%
Ratio	1.43
Hispanic Status	
Non-Hispanic	2.2%[NS]
Hispanic	1.8%
Ratio	1.22
Age	
18-24	5.2%
25-34	4.4%
35-44	2.6%
45-54	1.4%
55-64	0.8%
Ratio of 18-24 to 55-64	6.50
Education	
Less than High School	1.5%[NS]
High School Graduate	1.9%[NS]
Some College	2.9%
College Graduate	2.5%[NS]
Graduate School	1.1%
Ratio of Less than High School to Graduate School	1.36

Notes: Reference category in italics.
[NS]Indicates no significant difference in the probability of job entrances between categories of a two category variable or not significantly different from the reference category in a variable with three or more categories.
[a]Other demographic variables in model include marital status, household type, region of country, and residential environment.
Source: Authors' analysis of the March Supplement to the Current Population Survey, 1993-1996.

tion—for persons in all but one education level, those with some college.

Whites with disabilities were 43 percent more likely to enter jobs than their nonwhite counterparts (a difference that is statistically significant) and age was negatively significant, and monotonically related to the probability of job entrance. Persons with disabilities age 18-24 were 6.5 times more likely to enter jobs than those in the reference category, persons 55-64 years of age. Thus, the strongest demographic gradient n the probability of job entrance among persons with disabilities were due to age and race.

Owing to the large sample size of *persons without disabilities* in the CPS, relatively small gradients in the probability of job entrance reached statistical significance (Table 4). For example, whites without disabilities were only 1.15 times more likely than their nonwhite counterparts, and non-Hispanics without disabilities were only 1.16 times more likely than Hispanics without disabilities to enter jobs, yet both relationships were statistically significant. However, the issue of statistical significance aside, there were strong gender, age and educational

Table 4. Probability of Job Entrance for Persons without Disabilities, by Selected Demographic Characteristics, with Adjustment for Demographic and Work Characteristics[a], U.S. Average for 1993-1996

Demographic Characteristic	Persons without Disabilities
Gender	
Male	14.1%
Female	7.6%
Ratio	1.86
Race	
White	9.5%
Nonwhite	8.3%
Ratio	1.15
Hispanic Status	
Non-Hispanic	10.3%
Hispanic	8.9%
Ratio	1.16
Age	
18-24	12.7%
25-34	11.0%
35-44	10.0%
45-54	8.3%
55-64	3.0%
Ratio of 18-24 vs. 55-64	4.23
Education	
Less than High School	6.8%
High School Graduate	10.1%
Some College	9.6%
College Graduate	11.5%
Graduate School	13.3%
Ratio of Less than High School vs. Graduate School	.51

Notes: Reference category in italics.
[a]Other demographic variables in model include marital status, household type, region of country, and residential environment.
Source: Authors' analysis of the March Supplement to the Current Population Survey, 1993-1996.

gradients in the probability of job entrance among persons without disabilities, with men being almost twice as likely to enter jobs as women, those 18-24 years of age more than four times as likely to do so as persons 55-64 and persons with less than a high school education having only about half the probability of job entrance as those with at least some graduate school.

Table 5. Probability of Maintaining Jobs among Persons with Disabilities, by Selected Demographic Characteristics, with Adjustment for Demographic and Work Characteristics[a], U.S. Average for 1993-199

Demographic Characteristic	Persons with Disabilities
Gender	
Male	59.3%
Female	54.4%
Ratio	1.09
Race	
White	57.8%
Nonwhite	51.8%
Ratio	1.12
Hispanic Status	
Non-Hispanic	57.3%[NS]
Hispanic	53.4%
Ratio	1.07
Age	
18-24	52.3%[NS]
25-34	57.2%
35-44	59.4%
45-54	59.2%
55-64	52.5%
Ratio of 18-24 vs. 55-64	1.00
Education	
Less than High School	52.6%[NS]
High School Graduate	58.1%[NS]
Some College	58.4%[NS]
College Graduate	60.6%
Graduate School	55.0%
Ratio of Less than High School to Graduate School	.96

Notes: Reference category in italics.
[NS]Indicates no significant difference in the probability of maintaining jobs between categories of a two category variable or not significantly different from the reference category in a variable with three or more categories.
[a]Other demographic variables in model include marital status, household type, region of country, and residential environment.
Work variables include occupation, industry, size of firm, work hours, employee health and retirement benefits, individual and household earnings, and household nonearned income.
Source: Authors' analysis of the March Supplement to the Current Population Survey, 1993-1996.

Recall from Table 3 that the probability of job entrance peaked among persons with disabilities who had attended some college, then declined. Among persons without disabilities there was a clear break in the probability of job entrance between those who completed less than high school versus those with at least a high school education.

Table 6. Probability of Maintaining Jobs among Persons without Disabilities, by Selected Demographic Characteristics, with Adjustment for Demographic and Work Characteristics[a], U.S. Average for 1993-1996

Demographic Characteristic	Persons without Disabilities
Gender	
Male	85.4%
Female	84.5%
Ratio	1.01
Race	
White	85.2%
Nonwhite	83.5%
Ratio	1.02
Hispanic Status	
Non-Hispanic	86.4%
Hispanic	84.7%
Ratio	1.02
Age	
18-24	81.2%[NS]
25-34	85.7%
35-44	87.1%
45-54	87.4%
55-64	81.1%
Ratio of 18-24 vs. 55-64	1.00
Education	
Less than High School	83.2%
High School Graduate	86.2%
Some College	84.1%[NS]
College Graduate	85.8%
Graduate School	84.5%
Ratio of Less than High School to Graduate School	.99

Notes: Reference category in italics.
[NS]Indicates no significant difference in the probability of maintaining jobs between categories of a two category variable or not significantly different from the reference category in a variable with three or more categories.
[a]Other demographic variables in model include marital status, household type, region of country, and residential environment.
Work variables include occupation, industry, size of firm, work hours, employee health and retirement benefits, individual and household earnings, and household non-earned income.
Source: Authors' analysis of the March Supplement to the Current Population Survey, 1993-1996.

In Table 5, we report the impact of demographic characteristics on the probability of maintaining jobs *among persons with disabilities*. From the results, it is apparent that men with disabilities were slightly more likely to maintain jobs than such women and that whites were slightly more likely to do so than non-whites (both relationships, though weak, were statistically significant). Similarly, neither age nor educational level had a strong impact on the probability of maintaining jobs. Thus, the probability of maintaining jobs rose from just over 52 percent among those 18-24 to just under 60 percent among those 35-44 and 45-54, before declining to about 52 percent among persons with disabilities ages 55-64. Meanwhile, the probability increased from slightly more than 52 percent among those with less than a high school education to just over 60 percent among those who graduated from college, before declining to 55 percent among those with at least some graduate school.

Demographic gradients in the probability of maintaining employment were even smaller *among persons without disabilities* (Table 6) than among persons with them. Thus, among persons without disabilities, men and women, whites and nonwhites, Hispanics and non-Hispanics, and persons with various levels of education experienced nearly identical probabilities of maintaining jobs. Persons without disabilities ages 18-24 and 55-64 were less likely to maintain jobs than those in the middle age ranges, but, as among persons with disabilities, the impact of age on the probability of maintaining jobs was relatively slight.

The labor market literature provides evidence of strong demographic gradients in labor force participation rates; there is also evidence of strong gradients in such rates among persons with disabilities (Yelin and Katz 1994b). The results in Tables 3 through 6, however, suggest that, although several demographic characteristics do matter in determining the probability of entering and maintaining jobs for both persons with and without disabilities, the impacts are not as strong as in determining overall labor force participation rates.

In Tables 7 and 8, we evaluate the impact of work characteristics on the probability of maintaining jobs among persons with and without disabilities. Among persons with disabilities (Table 7), there were differences by occupation in the probability of maintaining jobs, but the pattern of impacts was not systematic. Occupations with a relatively low probability of maintaining jobs included such disparate ones as operatives (the backbone of manufacturing) and executive/professionals, while occupations with a relatively high probability of maintaining

Table 7. Probability of Maintaining Jobs, among Persons with Disabilities, by Work Characteristics, with Adjustment for Demographic and Work Characteristics[a], U.S. Average for 1993-1996

Work Characteristic	Persons with Disabilities
Occupation	
Executive/Professional	54.7%
Technical/Sales	57.2%
Administrative	62.1%
Service	55.8%
Crafts	59.0%
Operatives	52.1%
Transportation	61.6%
Laborers	56.2%
Ratio Lowest to Highest	.84
Industry	
Agric., Mining, Construction	44.7%
Manufacturing	52.8%
Transportation, Communication, Utilities	51.0%
Wholesale/Retail	60.5%
Finance, Insurance, Real Estate	55.3%
Service	57.9%
Professional Service	62.8%
Government	54.6%
Ratio Lowest to Highest	.71
Usual Hours of Work	
20 Hours Per Week	59.5%
40 Hours Per Week	56.1%
Ratio	1.06
Number of Employees in Firm	
< 10	65.8%
10-24	55.3%
25-99	58.3%
100-499	55.0%
500-999	54.2%
> 1000	50.2%
Ratio < 10 vs. > 1000	1.31
Health Insurance	
Employer Provides	59.6%
None Provided	55.2%
Ratio	1.08
Pension Plan	
Employer Provides and Employee Utilizes	57.9%
None Provided	56.4%
Ratio	1.03
Individual's Earnings	
10th Percentile	48.4%
Median	61.9%
90th Percentile	81.7%
Ratio of 10th vs. 90th Percentile	.59

(continued)

Table 7. (Continued)

Work Characteristic	Persons with Disabilities
Earnings of Remainder of Household	
10th Percentile	56.2%
Median	56.9%
90th Percentile	58.5%
Ratio of 10th vs. 90th Percentile	.96
Household Income Other than Earnings	
10th Percentile	58.5%
Median	56.5%
90th Percentile	52.5%
Ratio of 10th vs. 90th Percentile	1.11

Notes: Reference category in italics.
[a]Demographic variables in model include age, gender, race, ethnicity, marital status, household type, education, region of country, and residential environment.
Source: Authors' analysis of the March Supplement to the Current Population Survey, 1993-1996.

employment ran the gamut from administrative to transportation positions.

Differences among industries in the probability that persons with disabilities would maintain jobs reflect both the kind of work being done and the relative change in share of employment. The extractive (agriculture, mining and construction) and goods producing (manufacturing, transportation, communications and utilities) sectors had relatively low rates of job maintenance, while the wholesale/retail, service and professional services industries had relatively high rates. The latter have been experiencing rapid growth, while the former have had a declining share of employment, suggesting that the success of an industry may affect the probability of maintaining employment among persons with disabilities.

Interestingly, rates of job maintenance in government among persons with disabilities were not as high as in several other industries. Owing to Section 504 of the Rehabilitation Act, which banned discrimination in employment for persons with disabilities within government (and its contractors) considerably in advance of the passage of the Americans with Disabilities Act, one might expect that the probability of maintaining employment would be higher in this sector.

In addition to differences by occupation and industry in the probability of maintaining jobs among persons with disabilities, we evaluate several other aspects of work. Those working 20 hours per week as opposed to 40 were slightly more likely to continue working. Persons with disabilities in firms of fewer than 10 workers were 1.31 times as likely to maintain jobs as those in firms of more than 1,000 workers.

Table 8. Probability of Maintaining Jobs, among Persons without Disabilities, by Work Characteristics, with Adjustment for Demographic and Work Characteristics[a], U.S. Average for 1993-1996

Work Characteristic	Persons without Disabilities
Occupation	
Executive/Professional	85.4%
Technical/Sales	85.3%[NS]
Administrative	87.0%
Service	85.8%
Crafts	86.1%
Operatives	85.1%[NS]
Transportation	84.4%
Laborers	83.6%
Ratio Lowest to Highest	.96
Industry	
Agric., Mining, Construction	78.7%
Manufacturing	84.9%
Transportation, Communication, Utilities	84.1%
Wholesale/Retail	86.2%
Finance, Insurance, Real Estate	87.1%[NS]
Service	83.9%
Professional Service	83.9%
Government	84.3%
Ratio Lowest to Highest	.90
Usual Hours of Work	
20 Hours Per Week	82.9%
40 Hours Per Week	85.3%
Ratio	.97
Number of Employees in Firm	
< 10	87.3%
10-24	85.1%
25-99	84.8%
100-499	84.3%
500-999	83.4%[NS]
> 1000	83.4%
Ratio < 10 vs. > 1000	1.05
Health Insurance	
Employer Provides	88.8%
None Provided	81.9%
Ratio	1.08
Pension Plan	
Employer Provides and Employee Utilizes	86.7%
None Provided	84.2%
Ratio	1.03
Individual's Earnings	
10th Percentile	76.1%
Median	85.3%
90th Percentile	94.5%
Ratio of 10th vs. 90th Percentile	.81

(continued)

Table 8. (Continued)

Work Characteristic	Persons with Disabilities
Earnings of Remainder of Household	
10th Percentile	85.7%
Median	85.2%
90th Percentile	84.0%
Ratio of 10th vs. 90th Percentile	1.02
Household Income Other than Earnings	
10th Percentile	86.5%
Median	84.5%
90th Percentile	79.8%
Ratio of 10th vs. 90th Percentile	1.08

Notes: Reference category in italics.
  ^{NS}Indicates no significant difference in the probability of maintaining jobs between categories of a two category variable or not significantly different from the reference category in a variable with three or more categories.
  ^{a}Demographic variables in model include age, gender, race, ethnicity, marital status, household type, education, region of country, and residential environment.
Source: Authors' analysis of the March Supplement to the Current Population Survey, 1993-1996.

addition, there were slight, albeit significant, effects of having health insurance through one's employer (or union) and of having and utilizing a pension plan. Perhaps reflecting the kinds of jobs with low pay, perhaps reflecting a disincentive to work, persons with disabilities with earnings in the bottom decile were only about 60 percent as likely to maintain jobs as such persons with earnings in the 90th percentile. Interestingly, the earnings of the remainder of the household's members had almost no effect on the probability of maintaining jobs (those in the tenth percentile were 96 percent as likely as those in the 90th) nor did the amount of household income other than earnings (those in the tenth percentile were 1.11 times as likely to maintain jobs as those in the 90th).

Occupational and industrial differences in the probability of maintaining jobs were smaller among persons without disabilities (Table 7) than among those with (Table 8). Among persons without disabilities, those in the occupation with the lowest probability of maintaining employment—laborers—were 96 percent as likely to continue working as those in the highest—administrative workers. Similarly, whereas among persons with disabilities those in the industry with the lowest probability of maintaining jobs were only 71 percent as likely to do so as those in the highest, among persons without disabilities the analogous ratio was .90.

Among persons without disabilities, there were only small differences in the probability of maintaining jobs by usual hours of work, number of employees in the firm and health insurance and pension status. The probability of maintaining jobs among persons without disabilities with individual earnings at the tenth percentile was only 81 percent of those with individual earnings in the 90th percentile. The difference was smaller than among persons with disabilities, where those in the tenth percentile of earnings were only 59 percent as likely to maintain employment.

Similar to persons with disabilities, there were only slight differences among persons without disabilities in the probability of maintaining jobs as a function of differences in earnings of the remaining household members and in total household income other than earnings.

One method that has been suggested to assist persons with disabilities in maintaining jobs is to retrain them for occupations that they have the functional capacity to perform and that will position them well in the contemporary economy. An additional strategy is to pinpoint growth industries which may have a greater need for workers, including those with disabilities. To evaluate these strategies, we estimated models which included a core of demographic factors with industries alone, with occupations alone and with both (results not in tables).

We found that the addition of either industries and occupations to the regression with demographic factors alone increased the explanatory power of the models. However, the increment associated with the addition of industries was greater, suggesting that the strategy of hitching a ride on successful industries may improve the probability of job maintenance to a greater extent. Of course, retraining for specific occupations and then focusing on industries with the greatest growth potential may make the most sense, since the model with both occupations and industries performed better than the models with either set of variables alone.

Interestingly, when we estimated models of the probability of job maintenance among persons without disabilities, the increment in explanatory power associated with occupations was greater than that associated with industries, exactly the opposite of the situation among persons with disabilities. This suggests the hypothesis that persons without disabilities are not as dependent on the fates of industries to maintain their jobs as persons with disabilities. Armed with their other characteristics, including an occupation, they may be better able to buffer the impact of being in an industry with slow growth potential. Consistent with evidence from other studies showing that persons with disabilities may be part of a contingent labor force more subject to changes in the economic climate than persons without disabilities (Yelin 1992), the data reported here indicate that it is important that such persons be located in industries destined to have substantial growth. Otherwise, it may not be possible for persons with disabilities to buck the overall trend in which high proportions are forced to leave work.

SUMMARY AND CONCLUSIONS

The present paper reports the continuation of three disturbing trends: persons with disabilities are much less likely to be employed at any one time, if unemployed they are much less likely to enter jobs and if employed they are much less likely to maintain jobs than persons without disabilities.

We have found that differences between persons with disabilities and without in demographic and work characteristics account for a substantial fraction of the gap in their employment rates; a significant, albeit smaller, fraction of the difference in their ability to maintain jobs they already hold; and almost none of the difference in the ability to gain entry to new jobs. The implication of the latter finding is that disability itself, rather than differences in the characteristics of persons with and without disabilities, probably accounts for the low rate of job entry.

However, when evaluating characteristics of persons with disabilities that affect the probability of entering jobs, we found that those 18-24 years of age were more than six times as likely to enter jobs as those 55-64. One troubling finding is that whites with disabilities were 40 percent more likely to enter jobs when unemployed than nonwhites, after taking into account all other differences between the two groups. Thus, race would appear to accentuate problems faced by persons with disabilities (and vice versa).

The results with respect to race and age suggest that the equal employment provisions of civil rights legislation must be enforced on the basis of all of an individual's characteristics, rather than partitioning enforcement into separate initiatives. In addition, it would appear that getting persons with disabilities established in jobs at an early age is essential, because the probability of obtaining jobs for older workers declines precipitously. This strategy would not, however, make a difference for the many people whose disabilities stem from chronic conditions with onset in middle age or later.

Given the low rate of employment among persons with disabilities, using the CPS definition, such persons are less than one-third as likely to be employed) increasing the probability of job entrances must remain the central goal of employment policy. The research reported here provides no easy strategies to accomplish this goal. Other than age and race, there were no strong correlates of job entry.

In the short run, increasing the number of persons with disabilities who maintain jobs is a more attainable goal, not only because the probability of maintaining jobs is far greater than the probability of entering them, but also because a number of work characteristics had a statistically significant impact on the probability of maintaining jobs. These include the individual's occupation and industry and the size of the firm (which had a relatively strong impact on the probability of maintaining jobs), as well as the number of work hours, and whether the employer or union provided health insurance and pension plans (which had significant, although relatively weak effects).

The industries of persons with disabilities profoundly alter the probability of employment, even after taking all other demographic and work characteristics into account. Such high-growth employment sectors as professional services and wholesale/retail trade improve the odds of maintaining jobs, while low-growth sectors like agriculture, mining and construction and manufacturing worsen them. That the industry matters for persons with a disability has important implications for those currently working because there is evidence that it is easier to switch to new jobs while employed than to gain entrance to any job when unemployed (Yelin and Katz 1994a). This suggests that it may be prudent for persons to attempt to move to high-growth sectors as soon as possible after onset of a potentially disabling condition.

However, this finding also has implications for those who are not currently employed. Although entry into jobs among persons with disabilities is relatively rare, it is far less so for those under age 35 and for those with at least a high school education. To make sure that employment entrances are more than temporary, persons with disabilities must focus their job searches on industries with solid growth potential. Far more than persons without disabilities, the chance of maintaining employment for persons with disabilities is tied to the prospects of the industries in which they work. Given low rates of job entry and low overall employment rates at the present time, it is important not to squander any opportunity for job maintenance. Job referrals and enforcement of the equal employment provisions of the Americans with Disabilities Act through the legal system must be focused on gaining access to high-growth sectors of the economy.

REFERENCES

Adler, N., and K. Matthews. 1994. "Health Psychology: Why Do Some People Get Sick and Some Stay Well?" *Annual Review of Psychology* 45: 229-259.
Berkowitz, M., W. Johnson, and E. Murphy. 1976. *Public Policy Toward Disability.* New York: Praeger.
Blanc, P., M. Jones, C. Besson et al. 1993. "Work Disability Among Adults with Asthma." *Chest* 104: 1371-1377.
Brandt, E., and A. Pope. 1998. *Enabling America*. Washington, DC: National Academy of Sciences Press.
Jencks, C., L. Perman, and L. Rainwater. 1988. "What Is a Good Job: A New Measure of Labor Market Success." *American Journal of Sociology*, 93: 1322-1357.
Jones, N. 1991. "Essential Requirements of the Act: A Short History and Overview." In *The Americans with Disabilities Act: From Policy to Practice*, edited by J. West. New York: Milbank Fund.
LaPlante, M., and D. Carlson. 1996. "Disability in the United States: Prevalence and Causes, 1992." *Disability Statistics Report (7)*. Washington, DC: U.S. Department of Education, National Institute on Disability and Rehabilitation Research.
Levitan, S., and R. Taggart. 1977. *Jobs for the Disabled*. Baltimore, MD: Johns Hopkins Press.
Murphy, L. 1991. "Job Dimensions Associated with Severe Disability Due to Cardiovascular Disease." *Journal of Clinical Epidemiology* 44: 155-166.
Polivka, A., and J. Rothgeb. 1993. "Redesigning the Questionnaire." *Monthly Labor Review* 116: 10-29.
Pope, A., and A. Tarlov (eds.). 1991. *Disability in America*. Washington, DC: National Academy of Sciences Press.
Reisine, S., K. Grady, and C. Goodenow. 1989. "Work Disability Among Women with Rheumatoid Arthritis." *Arthritis and Rheumatism* 32: 538-543.
Stapleton, D., B. Barnow, K. Coleman et al. 1994. "Labor Market Conditions, Socioeconomic Factors, and the Growth of Applications and Awards for SSDI and SSI Disability Benefits." Report Prepared for the Office of the Assistant Secretary for Planning and Evaluation, Department of Health and Human Services.
Trupin, L., D. Sebesta, and E.Yelin. 1997. "Racial Disparity in Employment Among Persons with Disabilities, 1990-1994." Paper presented at American Public Health Association Meetings, November 11.
──────. (in press). "Transitions in Employment and Disability Among Persons 51-61 Years of Age." *Disability Statistics Report (15)*. Washington, DC: U.S. Department of Education, National Institute on Disability and Rehabilitation Research.

U.S. Bureau of the Census. 1993. *Current Population Survey*, March 1993.
Wilkinson, R. 1996. *Unhealthy Societies: The Afflictions of Inequality.* London: Routledge.
Yelin, E. 1992. *Disability and the Displaced Worker.* New Brunswick, NJ: Rutgers University Press.
_____. 1996. "The Labor Market and Persons With and Without Disabilities." Paper Prepared for Social Security Administration Office of Disability and National Institute on Disability and Rehabilitation Research.
Yelin, E., C. Henke, W. Epstein. 1987. "The Work Dynamics of the Person with Rheumatoid Arthritis." *Arthritis and Rheumatism* 30: 507-512.
Yelin, E., and P. Katz. 1994a. "Making Work More Central to Work Disability Policy." *Milbank Memorial Fund Quarterly* 72: 593-619.
_____. 1994b. "Labor Force Trends of Persons With and Without Disabilities." *Monthly Labor Review* 117: 36-42.
Yelin, E., M. Nevitt, and W. Epstein. 1980. "Toward an Epidemiology of Work Disability." *Milbank Memorial Fund Quarterly* 58: 386-415.

RESEARCH ON HEALTH CARE EXPERIENCES OF PEOPLE WITH DISABILITIES
EXPLORING THE SOCIAL ORGANIZATION OF SERVICE DELIVERY

Marie L. Campbell

ABSTRACT

This paper discusses the conceptual framework of a community-based, participatory, research project in Victoria, BC, Canada, in which people with disabilities and health care providers work together to understand the health care experiences of people with disabilities. Learning together is assumed to be a useful precursor to taking effective action, in the context of explicitly more "inclusive" health care planning (a goal of local health care reform). The paper argues that to offer useful insights for action by and for people with disabilities who are health care clients, the social organization of their actual experiences needs to be explored and critically analyzed. To do this, Dorothy Smith's (1987, 1990) institutional ethnography is employed and its use explained in the paper.

INTRODUCTION

People with disabilities have long-standing concerns about their health services that in Canada, as elsewhere, have been expressed in various lobbying and advocacy actions. Now, a major restructuring of health care is affecting both the services and the way that current claims and complaints are being dealt with. This paper addresses the intersection of these state-sponsored changes with a participatory, action-oriented research program called *Project Inter-Seed,* on health care for people with disabilities. In doing so, it makes an argument about how the research is conceptualized. The research uses institutional ethnography developed by Dorothy Smith (1987, 1990a and 1990b) to track the local conditions and practices of providing and receiving health services, not as discrete phenomena, but as socially organized, interactive ruling practices.

Institutional ethnography produces knowledge that makes the normally invisible social organization of people's daily lives available as a knowledge resource they can use to help transform it. While institutional ethnography has feminist roots, this method is a tool for activism that anybody can use.

I have found Anna Yeatman's (1994) theoretical proposals regarding the operation of the state under postmodern conditions useful in thinking through the context of this research. Especially relevant to our[1] research is her account of the role of emancipatory movements in provoking changes in the modern state. Her insights about the so-called performativity of the state are helpful to my analysis of the health care system that is under reform in British Columbia. Performativity, Yeatman says, has been substituted for paternalism as a basis of the state's control functions, as it responds to challenges from emancipatory groups.

"Performativity has the singular virtue of supplying a meta-discourse for public policy" (Yeatman 1994, p. 110). As a principle of governance, it establishes strictly functional relations between a state and its inside and outside environments and it does so in a climate where "the state itself is having to act more and more like a market player that shapes its policies to promote, control, and maximize returns from market forces in an international setting" (Cerny 1990, p. 230). It is apparent that within a meta-discourse of this kind, attention to efficiency and effectiveness of state services would be emphasized to further the state's competitive positioning. That is one tendency of health care reform.

But Yeatman also recognizes a contradictory dynamic arising from the activity of previously excluded groups. Claims on the health care system, for instance, express demands for inclusion being made by people with disabilities and establish a democratizing pressure in opposition to performativity. Yeatman argues that as the modern state encounters resistance to mainstream policy and practices from marginalized groups and individuals, its legitimacy is challenged. The contemporary state, she suggests, is constituted of emergent practices whereby state authority is reasserted in new ways that attempt to address, respond to and contain these challenges. This kind of thinking needs to be included in conceptualizing the experiences of people with disabilities within Canada's public health care system.

I bring to this research an ontological assumption that social life is produced in social interaction and therefore the study needs to be of people's actual practices. This may turn out to include performativity-inspired state management practices. But performativity is a theoretical concept that for Yeatman has broad implications. In borrowing the use of this concept for my own purposes, I am conceptualizing performativity as administrative practices, enacted by people (in this case in public health care) and concerted by means of various technological, thus material, processes[2]. These health care practices are the focus of analysis and of the project's change-oriented activities. To that end, I have designed an institutional ethnography to identify how the interaction between providers and people with disabilities is co-constructed and what, in specific instances, are its everyday outcomes for people with disabilities.

This paper is an "in progress" account of the conceptual framing of the project, based not just on the research design, but on work done in the preliminary fieldwork phase. The argument advanced is that to offer useful insights for action by and for people with disabilities who are health care clients, the *social organization of their actual experiences* needs to be explored. Only then can "reformed" health care practices, such as consultation with community groups and health care goals constructed in relation to indicators "found" in program-client interactions, be properly understood. Otherwise, the experiences of clients with disabilities disappear into the "performative" exercise of managing health care accountably, which some might claim is more related to establishing the state's competitive positioning than to improving people's health.

REFORMING HEALTH CARE IN BRITISH COLUMBIA (IN THE CAPITAL HEALTH REGION, VICTORIA, BC)

The Canadian health care system is undergoing "arguably ... the most radical restructuring ... since (the inception of Canada's national medical insurance legislation) with far-reaching implications for governments, citizens, physicians, hospitals and other interest groups" (Lomas et al. 1997, p. 372). Canadian health reform draws inspiration from the World Health Organization's work which has popularized broader definitions of health than simply absence of illness, and has opened up to debate the question: "What produces health?" This question challenges the hegemony of biomedicine and allows health policy to address "intersectoral activities having to do with sanitation and clean water, nutrition and healthy eating habits, physical exercise, employment, education, housing and so on" (Saltman 1997, p. 444).

More recently, the notion of control over important aspects of one's life has been proposed as a major determinant of health (Evans et al. 1995). In spite of the work of WHO in keeping notions of "health for all"—access and equity issues—in the forefront of health planning, Canada's health care reform still expresses and maintains the historical dominance of acute care and thus of biomedicine. One influential critique, written from a political economy perspective, argues that health care reforms in Canada are "based mainly on the medical model and are chiefly designed to change who pays" (Armstrong and Armstrong 1996, p. 10). The Armstrongs identify the context of health care reforms as "a new trade agreement that reflects and reinforces an emphasis on efficiency and effectiveness defined in for-profit terms" (p. 226), and a political climate that is fixated on a debt and deficit crisis "justifying both cuts in (social) spending and a move towards a free market system" (p.7) in health care. They write what many Canadians are coming to fear about health care reform: "that opportunities to remain healthy and receive useful care are declining ..." (and they predict) "increasing inequality, fewer choices and less skilled care" (pp. 225-226).

Most observers see finance and expenditure-related matters, including integration of hospital-based and community services, as the dominant motivation in health reforms (Saltman 1997; Lomas 1997). As the Armstrongs' comments suggest, more is at stake than cost reductions. Lomas (1997, p. 818) proposes that under health care reform "the preservation of private domains is expected to give way to public objectives." Public objectives as Yeatman's discussion of performativity suggests, can mean an entirely different conception of the role of the state in people's lives than under a welfare state, for instance.

In British Columbia, following the report of a royal commission[3] that investigated the operation of the provincial health care system, the government announced its reform policy "New Directions for a Healthy British Columbia" in 1993. That policy mandated a broader definition of health that would lead to the development (through public consultation) of "health goals" and to monitoring progress in meeting them. It encouraged greater participation by citizens in their own health and offered a variety of information supports. It laid plans for bringing health services closer to home through expanding community care and providing local management of health services (regionalization). It announced support for health care providers while hospital and community health services were being integrated (such as labor agreements that prevented layoffs for a defined period of time).

Along with decentralizing management of health care, the reform plan put increased emphasis on new methods of accountability for the services provided, based on outcomes measures (British Columbia, 1993). In regard to the latter, it is important to note that at the same time as the organization of health care was being reformed, the provincial government implemented a new accountability framework to cover, not just health services, but the whole public sector (British Columbia, 1996). Explaining the paradigm shift in public accountability being undertaken, a government handbook says:

> government reporting and management continue to focus heavily on resources, activities and compliance with rules. While these are important, more attention should be given to organizational and program results, and performance with respect to the way business is conducted. We believe that government process in British Columbia will be enhanced if legislators and government focus more on results: what is working and what is not and where increasingly limited resources can be utilized in the most relevant, economic, and effective manner (p. 17).

To this end, a "performance management" system, whose central feature is outcomes recording and measurement, is being introduced into all departments and agencies of government. This is the context in which the study of health care for people with disabilities is being conducted. Later in the paper, we will see some evidence of how the public administration of British Columbia's health care system is being brought into closer coherence to policies for successful management of government. To advance the cause of adequate attention to health care for people with disabilities, it is argued here that their standpoint should not be submerged into generalized accountability practices.

THE STUDY

Project Inter-Seed: Learning from the health care experiences of people with disabilities is a participatory and action-oriented study being conducted in Victoria, BC, Canada. The project is formally set up as a partnership between a community agency, the South Vancouver Island Resource Centre for Independent Living and the University of Victoria, where the author and principal investigator is a faculty member. The research is funded by two scholarly granting bodies and a local philanthropic foundation[4].

Participants with disabilities as well as some health care providers are being trained as researchers and paid. In the case of employed health care workers, when they work on the project they are "replaced" in their jobs from project funds. These participant researchers will carry much of the responsibility in the latter stages of the project to design and lead actions based on our developing findings. Their learning from the research (as well as the knowledge they bring to it) is crucial to the success of the project.

The knowledge objective of the project is to explicate what are the common health care experiences of people with disabilities by exploring the social organization of the delivery of services. In phase one, narratives of participants' experiences with health care were collected[5].

The main focus of the second phase analysis is a "taking stock"[6] and we have decided to focus analytic attention on the operation of one health care program—the region's Home Support program[7]. Taking stock here means detailed observations backed up by interviews and documentary analysis of formal, text-mediated decision points in the everyday work of providing health care services to people with disabilities. We expect to discover from this taking stock exercise how specific instances of health care provision or administration could give rise routinely to the stories being told. Having heard similar stories from a number of informants, we assume[8] that they may have a socially organized, not an idiosyncratic, basis and we go looking for such organizational features. While health care and home support are not necessarily the determining features of people's lives, in many cases it was around health care decisions that people with disabilities experienced either a broadening or a narrowing of their options for living their lives.

"PROBLEMATIZING" THE RELATION BETWEEN HEALTH AND HEALTH SERVICES FOR PEOPLE WITH DISABILITIES

Institutional ethnography's interest in exploring and explicating how experiences are socially organized directs research attention in a particular manner. Project Inter-Seed begins, for instance, not with a hypothesis but with people's actual experiences. This is a central feature of institutional ethnography. Volunteers were recruited to recount to researchers their stories about health care experiences.[9] The goal was to "problematize" the relation between what is already known (authoritatively) about the services provided and what actually happened to individuals with a disability. That preliminary data from stories provided the experiential basis for *what would be studied* in phase two: the social organization of specific health care interactions in other actual settings in the Capital Regional District.

The stories illuminated the question of what "health" is, to people with disabilities. This topic arose as the research team considered the critical framework that would guide our "taking stock" observations of the delivery of Home Support services. To keep the experiences of the participants at the center of the research, we needed to think about how their health and their services were actually related. We turned to the narratives. Carol (a pseudonym) says:

> health for me is not just centered on whether I'm physically able or not. For me, I think it's whether I am able to achieve the kinds of things I want to do in my life and what either facilitates or gets in the way of that. ... I'm healthiest when I'm not being kind of pushed in by things around me. Included in that is things happening in my body. ... so if my body is really bothering me a lot and I'm confined because of my body, or there's some kind of barrier that I'm up against, let's say

at the university or my insurance plan, then I'm feeling less healthy because all of a sudden everything is narrowed down for me. So, (health) is kind of about having a fair amount of room to play, but also having the support (I need).

This story and others pointed to the everyday significance for the health of people with disabilities of health care services being supportive of their autonomy. The concept of independence is a key plank in the framework of the disabilities movement and our research partner made sure that Independent Living tenets[10] are part of the research framework. In the discussion around independence in the research team, it became clear that the project's participant researchers wanted to separate the desire for autonomy from the view of independence as "standing alone"—a version of independence that they thought nondisabled people might hold. Their view of good health care services is that, besides being of therapeutic value in the medical sense, they should also contribute to building the capacity for persons living with a disability to exercise control in their life. And in addition, participant researchers insisted that in their personal lives they need to feel connected with and sometimes even dependent on other people.

Some further comments made by Carol touch on the intersection of relationships, health care services, and disabilities. As can be interpreted from Carol's words, her health and her health care needs can't be adequately understood by assessments of her physical capacities alone. She continues:

I'm dog tired. And I just get the basics done. My partner is away. I've been on my own since August and it's been hard because my home care was cut off two years ago because I'm not sick enough—which is such a joke. I can't do vacuuming. I don't do heavy housework at all. So he'd come home, once a month he'd spend the weekend cleaning and doing the heavy housework for me.

Carol found this unacceptable as a solution to her difficulties with heavy housework. She recognizes that relationships can break up when strained by the extra demands that come from being a caregiver to a partner with a disability. Destroying a relationship is an unacceptable "economy" in health services. It would be one more, very costly, narrowing of a person's life.

Like this participant, many other people with disabilities who participated in our inquiry were able to identify common instances when the health care they received was not as helpful as they thought it might or should have been. Besides "narrowing" a person's options, as above,

Research on Health Care

when services were not available to support the person's maximal functioning, we heard that getting services sometimes posed frustrating barriers, or that the services allocated may not have been appropriate.

A somewhat different perspective on the way that health services relate to "taking control of one's life" comes from Jessica, another participant with a disability. Jessica is a young, active woman who uses a wheelchair because of a spinal cord disability. She now lives alone in accessible housing, has a disability pension and engages enthusiastically in competitive wheelchair sports around which her social life is organized. But she wasn't always this autonomous. Earlier in her life, she lived in a hospital for several years and some of her caregivers thought she would always need to be in a long-term care facility. She values her independence highly and has struggled to take care of herself so that she can remain active in the world and "have a life."

The story that Jessica tells is about her medical insurance. Medical insurance is a universal program in Canada, federally funded by tax dollars and administered by provinces; it pays for certain amounts and kinds of health services available from professional practitioners on a fee-for-service basis. Jessica discusses a source of frustration about payment for physiotherapy treatments that she needs:

> Physio—well, you are only allowed twelve visits per year. Last year I had a lot of problems with spasticity in my back and this is still ongoing. I used up twelve treatments and I could not get ... any more ... that they would pay for. Massage is covered for twelve (treatments), so I jump from one twelve to another twelve. Like I went to twelve physio and twelve massage and by then I was at the end of the year and I could restart. It was just really frustrating because I know physio would have helped it much more than massage. Massage relaxes the muscles, but physio treats it more aggressively in different ways. I was told to go back for physio for my spasms, but I'll run into the same problem. I'll just be doing it once a week for twelve weeks and it will be finished. ... things just get started and then you have to stop. Why start something, because you're making progress and you have to stop that progress. Anybody who is on medical (insurance) gets twelve treatments, but we have a disability, which is more frustrating.

The rationale for limiting the amount of (expensive) service that any one citizen can claim appears to be some notion of equity. That is, although everyone's access to free services is limited to twelve treatments, everyone has equal opportunity to access those limited amounts of scarce resources. Jessica points out that this may be equal, but in her view, it is not appropriate. In addition to physio, Jessica is being treated pharmacologically for back spasms, with a very costly drug and drug pump. She is both anxious to be "drug-free" and finds physio helpful, but the rules prevent her from having a sustained program of physio. She finds this both irrational and annoying. It clearly limits her options for how she can manage her disability effectively.

Another participant, Mary, identifies how decision-making about home support services narrows her capacities instead of supporting her autonomy. Mary has arthritis. She has been telling the interviewer that her need for help at home varies because her health does. However, the rules about home support are inflexible. At the present time, her home support has been discontinued. To enter into renegotiation of her eligibility with each alteration in her health would be a burden, and she explains why:

> Now, for instance, if I was to go and get physiotherapy, which let's face it, with this knee I could probably do, guess what? I get a housekeeper again. Like, I mean that doesn't make any sense to me, but that's the way it works. If I'm getting service somewhere else through long-term care, then I can have my housekeeper back.
>
> **Interviewer:** And why wouldn't you do that now that you've got a sore knee?
>
> I don't know. Part of it I think is the hassle of it. I guess it's part of my health. It's kind of like an ongoing line that keeps shifting around how disabled am I? How ill am I? Right? It's a zone of negotiation, right, where I negotiate with myself to begin with about how much I'm able to do at any given point in time, right? Any moment, any hour, any day. Then along with that, it doesn't sit in isolation, it goes along with, like, what do I say to other people? like when people say "how are you?" Nobody wants to hear about my sore joints, right? So I don't generally tell people about them. But sometimes, like for me to talk to my long-term care assessor, I really like her a lot, (but) it's part of this whole thing, my reluctance is. I have to call (her) up and I have to go through the whole thing of ... I have to tell her how much money I'm making, I have to tell her this about my life and that about my life, and I basically have to hand over everything and somebody makes the decision whether I get the care or not. And you know what? Sometimes that's just too much of a hassle.
>
> **Interviewer:** And maybe it's not just a hassle, maybe what you're talking about is like giving up your identity as a competent... autonomous person.
>
> I agree. I agree a hundred percent with that. And that's the negotiation, because for me I go from almost sometimes absolute dependency to being quite independent and autonomous, right? Like sometimes I can't get out of bed. So that for me that's absolute dependency. Or sometimes I'm emotionally upset so badly that I can't talk to anybody. Well, that's pretty dependent too. So now I have to call (my long-term care assessor) and portray myself as ... . (B)ut she might tell me, sorry you make too much money, if they have to look at, you know what they use. So that kind of an invasion, like, ahh, forget it!

This is an eloquent presentation of the disempowering potential of a service meant to be supportive of the health of a person with a disability. Mary and others talked about managing the impression of themselves given to others, including health care personnel. Mary points out that to make herself fit eligibility criteria, she would have to undermine her own definition of herself as "doing OK" or "not being too disabled." Because her health fluctuates, she sees the project of explaining herself to establish eligibility for needed services as never-ending and, itself, disabling. She has opted to back away from getting the service to which she is actually entitled.

These insider views of service provision offer some sense of the lack of fit between "generic" rules and procedures and the health care needs of people with disabilities. They highlight, also, the links between health, health services and autonomy that these people with disabilities are able to make from their own experiences. Project Inter-Seed and its subsequent ethnographic data collection and institutional analysis attempts to make those kinds of links available and understandable in a more systematic, less "personal" form, as a legitimizing basis for action.

CONTESTATORY ACTION

The next step in our research is to document the actual interactions in which health policy, programs and everyday organizational-professional decisions accomplish this kind of not-quite-fitting or inappropriate and dangerous service provision. It is that kind of knowledge that will be the basis of change-oriented action directed toward state-sponsored service providing agencies. Yeatman (1994) argues that the challenges from emancipatory movements against the rational consensus that characterized the legitimacy of the modern state and its policies are being responded to in ways that constitute the contemporary state, not as a break with the past, but a revision of it.

To illustrate her argument she offers instances of feminist and aboriginal struggles against exclusionary practices, and against policies constructed on assumptions unacceptable to these groups. Yeatman's point about such challenges mounted by these contestatory groups is that a public rational consensus on policy holds as long as the society supporting it can be taken for granted as being homogeneous. A customary basis in Britishness and liberal ideas that has underpinned both

British democracy and its colonial replications is being undermined by immigration into settler societies like the Australia of which she writes (or, presumably, Canada). Likewise, the customary paternalistic basis of social policy has been challenged successfully by feminists in struggles over state responses to sexual assault, domestic violence and abortion issues, etc. In post-colonial struggles, property rights arising from a liberal political discourse are not accepted by aboriginals, who hold beliefs about the land supported by different grounds than those appealed to by settler/citizens.

When it comes to struggles over health care, the narratives already referred to in this paper open up for similar inquiry the accepted rightness of "generic" rules about health care provision and accountability. People with disabilities disrupt the cultural homogeneity of a population whose ability level is taken for granted. It is easy to see, for instance, that the idea of cultural homogeneity (of able-bodiedness) is disrupted by Jessica's and Mary's stories. It appears that the criteria in use for making routine decisions about service provision (if we take Jessica's story as an instance) are inappropriate. Rules and regulations made for the non-disabled, presumably not having included the disabled in their conceptualization, are being questioned.

Once disrupted, the legitimacy of taken for granted rules and procedures challenged, what is the course of action to be followed? Is it a free-for-all? Do the voices of people with disabilities take precedence over other contenders? Historical, custom-based claims and claimants do not just acquiesce in the face of competing claims. Does the state's authority collapse in the face of multiple voices? This is where the politics of difference meets the emergent practices of postmodern statecraft, according to (my reading of) Yeatman.

In this politically charged climate, emergent performative practices become the operative "solution", although never a final solution. Dispute resolution mechanisms have become very popular in British Columbia as an approach to resolving differences (Hume 1997). Yeatman seems to suggest that in the postmodern state, such approaches for working with "irresolvable differences" among contestants might be considered a successful way of negotiating temporary settlements (Yeatman 1994, p. 115). Yet, it is my understanding that dispute resolution treats contestants as equal, and multiple voices as equally valid, a performative notion that cancels power differentials in a manner that, or people with disabilities, could simply obliterate rather than respond to their different needs.

Project Inter-Seed's research is not attempting to make a case for contestatory action *from the narratives collected* in the first phase of the research. Although the stories are not idiosyncratic, as noted earlier, the experiences of the people interviewed are diverse. Where Yeatman speaks of the differences among members of emancipatory groups resulting in multiple perspectives, we note, in contrast, that people in the narratives we collected have a range of *experiences*. Our analysis is not of the subjective content of people's views based on their experiences, as would be the focus of Yeatman's interest (I suspect). We do, however, count on people's experiences methodologically, as material evidence of what actually happens out of specifically organized, and policy-based, action.

The purpose of the narratives is to identify the decision points (and other key areas) where differences arise between official policy and what people with disabilities think constitutes appropriate and health-inducing service provision. Such points are the focus of further inquiry because they offer entry into the policies and their bureaucratic and/or professional implementation. We recognize that decisions about eligibility, for instance, are made through policy-framed organizational processes. These official decisions may not be the same as a professional's own judgment, but frequently professionals administer the documentary technologies that, besides "hooking people up" to services, also determine what is eventually done and its accountability.[11]

Professionals execute these decision technologies expertly, and in ways that have some effect on what actually happens. Professionals may become committed to the official version for a variety of reasons, including enforcement procedures such as chart audits, outcome measures and performance evaluations. It is the everyday operation of all these organizational and text-based processes and the authorization of them as expressions of specific policies that our "taking stock" aims at.

If our contestatory actions are to alter official thinking and dislodge established procedures in favor of practices that work better for people with disabilities, we have to understand how and why things are done as they are now. In the terms that institutional ethnographers use, these are the *relations of ruling* that constitute people's lives as they experience them. Relations of ruling exist as everyday practices, not just as theoretical constructs. Or, put another way, relations of ruling exist only when enacted locally, as in this case in health services.

EXPLORING HEALTH SERVICES AS TEXT-MEDIATED RULING PRACTICES

Health care reform in British Columbia is generally taken for granted by citizens as a necessary and useful response to spiraling health care costs and system inefficiencies. Reform, in British Columbia, as in other parts of Canada, is bringing various kinds of organizational restructuring and devolution of authority, packaged as part of a better, more responsive, perhaps even more equitable, health care system (Capital Health Region Board 1997). But within both worker and client groups there is still a nagging concern as to how the quality of publicly funded health care as Canadians have known it can be protected, given the climate of fiscal restraint (Sibbald 1997; Barlow and Campbell 1995).

If the performativity principle were taken to be at the heart of health care reform, state action would become oriented to the competitive "business" among and beyond national states superseding responsiveness to citizens. But, as already mentioned, claims from emancipatory groups may also bring a democratizing challenge to the state. How is this contradictory dynamic being played out in British Columbia within health care reform policies and practices?

People with disabilities want to be part of the debate that is currently in progress about health care reform. Being heard and responded to requires a process for doing so. Reform of the health care system and restructuring of the local administration under a regional board has been the occasion for implementing some practices aimed at greater citizen involvement. Included are strategies for responding to the claims from those who have previously been marginalized. New consultative processes are being devised to draw such expressions into mainstream health planning. For instance, recent newspaper ads in Victoria, B.C., invited citizens to provide feedback on a range of services, as part of a comprehensive review of the Capital Health Region's programs. Health care reform emphasizes "community consultation" and the CHR has circulated a plan that includes "improving customer satisfaction ... [and] public confidence ... in services, improved community awareness

... [and devoting a] greater percentage of resources to health priorities identified by the community" (Capital Health Region 1997, p. 3).

Under the province's reformed policies, a superstructure of accountability for health goals is being built as this jurisdiction's methodology for managing health care (British Columbia 1997). New practices for inclusion of multiple voices are being used to construct both the goals and the indicators to test how the goals are being met. (British Columbia 1995, 1997). The newly formulated overall goal for regional health care programs is "to maintain and improve the health of British Columbians by enhancing quality of life and minimizing inequalities in health status" (British Columbia 1997, p. 6).

Under the reformed health care system, the plan is to make services accountable to yearly measurements of health status. This is a new departure in health governance in British Columbia.[12] Health goals are expected to "provide a basis for setting priorities, linking public policy decisions and investments to desired health outcomes, and helping to ensure accountability for those investments and outcomes..." (British Columbia 1995, p. 5). The Capital Health Region, the jurisdiction in which our research is being done, is setting up its programs in such a manner as to be able to report "outcomes" in relation to specified program goals. Each of the CHR's services is developing its operational plans with the new requirement of outcome measures built into implementation plans.

Outcomes measures offer official knowledge of "what is" and "what needs to be" about health care. The new approach to health administration could be seen as an "emergent practice" for solving some of the problems associated with legitimacy challenges. Yeatman's contention is that emergent practices for including the voices of previously excluded citizens shore up the state's authority by revitalizing its basis of legitimacy. Building health goals with documentary targets and indicators actually provides a more authoritative version of health care needs and how they are to be met, building in but superseding the multiplicity of voices and perspectives of claimants.

The decision framework (health goals) replaces unilateral decision making, on the basis of "taken for granted" consensual policy and practices. It replaces the favoring of any one group by choosing their claims over another. This kind of text mediation of knowing (that is, the technology, itself) frees knowledge from specific knowers and the action that it authorizes can then be seen as unbiased. Health care reform thus appears in the CHR's practices as building a decision framework legitimated not through custom and professional expertise, but by community consultation. (Indeed, it would be possible to read the development of information systems that relate program outcomes to health goals in the CHR as authorizing and legitimizing an authoritative voice, *through a collective process.*)

The use of such technology solves the administrative problem of a multiplicity of voices and irresolvable differences. But the standpoint represented in such a concept of "the problem" is a ruling standpoint. This problem of multiple voices is the problem as an administrator would know it. To people with disabilities, the problem is of being excluded, living with services that don't fit, etc. The differences between these positions we expect to be clarified in Project-InterSeed's analysis of health care provision.

For instance, text-mediated forms of decision-making are not new, but their use is still not well understood. Some discussion of how they work (and for what purposes) and an example are offered here. The utility of text-based decision making is taken for granted as a tool of effective public administration (see Larson 1997). But there is more to text-mediated forms of action than effective public administration. Texts, according to Smith, (1990b) are constituents of social relations, not just substantive messages. This is another defining feature of institutional ethnography. In health care, textual practices tie internal processes of service delivery to extended social relations among the state, professionals, managers and recipients of services. Text-mediated action is a form of power that bureaucracies exercise. Texts routinely used by care providers carry a definite form of organization of social relations. In this study, we want to explore how ruling is accomplished, textually, in the process of providing health services. From an example, (drawing on Miller 1997) that is provided next, it is possible to see how a text-mediated professional practice carries powerful and troubling meanings into a relationship between a service provider and service recipient. We will observe "how a text as petrified meaning structures a reader's interpretation, and how its meaning may be entered into succeeding phases of the relation" (Smith 1990, p. 223).

Rena Miller analyzed her own interaction with community workers during her husband's terminal illness, in which she was his main caregiver. She discovered from her analysis of texts (requested through Freedom of Information legislation from the health care agency involved), how the nurses and social workers with whom she interacted "saw" her and recorded their side of the interactions. Their work with organizational records not only structured the meetings that took place between them but, as was the purpose of the official recording, they structured Rena and her family into "problems" solvable by their organizational programs and themselves, the organization's professional employees.

The palliative care discourse, Miller's research disclosed, was the legitimizing source of the ideas expressed through programs and program interventions. The palliative care assessment procedures framed how Miller could be and was seen "officially"—how needy she was emotionally, financially, physically and so on. The assessment framed how she was responded to by program workers. As an individual, Miller felt "ruled" and misunderstood, even though she recognized "good intent" and even competence on the part of the workers. Her own views of her needs were disregarded. Those views did not fit the categories of the text (the assessment form), nor the organization's needs for how its employees' work was to be organized, coordinated and accounted for. The organizational texts of the palliative care program, as petrified meaning, structured the professionals' interpretations and hence dominated the successive phases of the "helping" relationship.

Miller's research is an example of the kind of inquiry that Smith recommends when she says:

> The actual practices ordering the daily relations that regulate contemporary advanced capitalist society, however conceptualized, can be subject to empirical inquiry, to ethnographic exploration, once texts are recognized as integral and "active" constituents. Uncovering texts as constituents of relations anchors research in the actual ways in which relations are organized and how they operate. (This enterprise can transform) our understanding of the nature of power when power is textually mediated (Smith 1990b, p. 224).

Research interest, in Project Inter-Seed's phase two, focuses on the everyday practices of delivery of Home Support services, but from the standpoint of people with disabilities. Unlike Miller's analysis, where she compared her own experiences and a journal she had kept against the official texts, our research won't be attempting to match observations back to an individual's particular circumstances. Our analytic interest is in discovering how the narratives (of people's health care experiences) could have happened as they did. Our observations of text-mediated service delivery will lead us to the connections that we need to make. The texts themselves, as Smith points out, don't "appear from nowhere" (1990b: p. 223).

For instance, if "outcomes" are to be generated from recording done by health care providers and intended for a health goals measurement process, the organization of that intended use will appear in the work with clients. It will structure successive phases of the relations, as Miller found was the case. The recording of her life as "problems" in a problem-oriented work plan dictated a line of action that the nurse carried out—a relation in which Miller's own experiences were rewritten so that the agency's organizational purposes could be served. As Project Inter-Seed explores the reformed health care practices, thinking about the new inclusive approaches to knowing authoritatively may help us see how the Capital Health Region's public administration shores up its legitimacy in the face of a variety of challenges. And we will want to discover how this plays out in the lives of people with disabilities.

Project Inter-Seed is building knowledge to support contestation (broadly speaking) of the health care system's ruling practices that people with disabilities identify as not in their health interests. Our "taking stock" will explore the construction by actual people (health care providers and people with disabilities and their families) of their everyday worlds, in which authority is being exercised in various ways (as official processes, rules, etc). This kind of research seeks to understand power and authority as socially organized practices of ruling in which organizational texts and text-mediated activities carry (official) meanings and thus legitimacy across geographic and temporal sites.

For example, taking stock of Home Support might be constituted as follows. A team would be put together from our research group[13] to make observations of staff interacting with clients, doing some everyday routine work such as "assessing" different clients with disabilities for eligibility for Home Support services. Our observations and interviews are conducted for the purpose of learning a setting as a socially organized course of action, and as social relations.[14] The stock-takers will already have mapped the formal procedures for the health intervention they are observing. They will have collected and read the recording and reporting texts the worker will be using. They will listen for special language use in the setting. Smith's words explain this process of inquiry:

A method inquiring into the social relations of the setting to make these (social relations) explicit in description must begin with the language uses of the setting, and with the speech of those who are members of the setting. ... As they tell (the researcher) what is going on, and as they explain, name and explicate ... their teaching is continually informed by their implicit knowledge of the organization of the setting. ... There is an already intimate linkage between the terms members use to describe settings to newcomers and the social organization of the setting (they describe) (Smith 1990b, p. 118).

This ethnographic attention to the health care workers as "local experts" is to capture the organization of their work in their implicit knowledge of the setting. Insiders know it implicitly and enact it routinely. Thus, as insiders, expert practitioners of the health care interactions will open up the social relations of ruling that are "in" their work and their talk. This will expose how their work links them and their clients into ruling relations and *if* it links to a metadiscourse of performativity, we should see *how*. Like the palliative care discourse that Miller found was structuring responses to her as "problems," the metadiscourse of performativity may be replicated in the local health care setting when health care workers take up their work to address issues of accountability, efficiency and costs. If so, its concepts will appear in policies and again in locally text-mediated practices.

CONCLUSIONS

That the principle of performativity may be guiding health care practices in British Columbia is an interesting theoretical proposal and one that carries strong intuitive appeal. It seems to match and elaborate the findings of others who have been researching health care reform. Yet, in this paper, I find that I must return, as we do in the research itself (in the "taking stock") to the actual individuals—the health care providers who are interacting with clients, the project participants, and their everyday lives. I return to the materiality of the experiences that is to be understood.

Institutional ethnography offers a "scientific" account to counter the organizational account of health needs and health provision. Science here means the production of an account that is faithful to people's experiences (that reflects back to the narratives and explicates how those experiences could have been organized to happen as they did: the research writing makes it possible for anyone to "see" the connections).

Instead of taking ruling for granted, our account will make ruling relations explicit. Taking stock of a set of Home Support program activities will elaborate the connections between (what happens to) clients and the organization's routine procedures. Mystery now shrouds the shortcomings of carefully crafted and well-intended plans and procedures, as they are realized in interactions of service delivery. Generic troubles, failures, inadequacies and misunderstanding plague service delivery and, if noted at all, often get blamed on individuals.

Our research targets organizational processes, assuming that processes are executed routinely and, for the most part, competently. The troubles are expected to be routinely constituted also, perhaps in relation to a ruling metadiscourse of efficiency (or other concern), certainly in relation to the view of nondisability as an accepted norm, around which service provision is conceptualized. Learning these connections between people's experiences and organizational processes is a step in learning how service provision can be reviewed and revised in ways that work better for people with disabilities. That is the long-term objective of this research. The analysis will spell out, in the detail of the specific instances analyzed, how routine organizational processes work for or against people. If Project Inter-Seed is successful, it will help people with disabilities in their struggles, alongside health care providers, within systems that in large part determine how they can live their lives integrated into, not excluded from, society.

NOTES

1. In this paper I speak of "our research" when I refer to the work done in Project Inter-Seed. When I speak of the paper's analysis, I speak of "my analysis." However, as I mention within the text of the paper, our research team includes participant researchers whose analytic work instructs my own thinking at times. Yet, as author of this paper, I take full responsibility for what it says. Sally Kimpson, a member of the research team and Tanis Doe, Advisory Committee member and liaison from the Resource Centre for Independent Living, both read and provided valuable comments on this paper.

2. It should be noted that this paper attempts to bring into closer alignment some divergent ontological positions, as carefully articulated by Mann and Kelley in *Gender and Society* (1997). Especially in relation to the question of materialism, Smith's view and my own seem opposed to the idealism of many postmodern writers. However, Yeatman's references to local everyday life and to policy seem to me to come very close to recognizing the materiality of the terms that she uses. This paper pushes the exploration of the social organization of such knowledge even further.

3. The British Columbia Royal Commission on Health Care and Costs (1991) introduced into the BC policy environment some important thinking about community involvement in health care and health planning that was being expressed in WHO and Canadian health policy documents in the 1980s (e.g., the Ottawa Charter). Here, analysis shifts to the policy framework for implementation that was introduced in BC in 1993 and that takes up the Royal Commissions recommendations, in specific ways.

4. Appreciation is expressed to the BC Health Research Foundation, the Social Sciences and Humanities Research Council of Canada/Human Resources Development, Canada and the Vancouver Foundation, for funding the research. My analytic work has been advanced by discussions in a research network funded by SSHRC Grant # 816-94-0003 "Understanding and Changing the Conditions of Caring Labour."

5. Fifty-five informants volunteered to be interviewed in phase one. They were mainly lower-income, white women of all ages with disabilities and white parents of disabled children. Data gathering in phase two has increased the number of men involved in the project, but not the representation of ethnic minorities, unfortunately. We have had one man and one non-European immigrant as participants with disabilities on the research team. Generalizability in institutional ethnography arises from demonstrating general features of how informants' experiences are organized. Thus, representative sampling is not carried out nor is it required for demonstration of the study's trustworthiness as to how certain experiences arise. There is more work to be done, however, around our lack of inclusiveness with regard to diverse populations.

6. This is the name that participants chose for this phase of the research.

7. Home Support in the Capital Health Region of BC falls within the provincially funded Long Term Care program. People assessed as eligible for publicly subsidized Home Support services get a certain number of hours of personal care and help with activities of daily living, through a number of private home support agencies that contract with the local health region.

8. This "assumption" is really a methodological tenet of institutional ethnography. It arises from a view of the world as socially and interactively constituted, through social relations—people acting in concert but not necessarily face to face, and this includes participating in the relations of ruling that dominate them (Smith 1987, 1990a and 1990b). Action is frequently concerted through texts in discursively organized practices. In assuming that research informants' health care experiences were socially organized, we rely on previous research on the social organization of health care, such as Campbell (1988, 1992, 1998)

9. Several reviewers have queried the propriety, in social science terms, of the kinds of questions and responses made by the researchers in the course of these conversations that we called "interviews." The attempt was for researchers to thoroughly understand what the volunteers wanted to tell us about their health and health care. Rather than being data from which we would develop explanations, these narratives showed us "what needs to be explained" subsequently. They helped us decide what other data collection and analysis were needed. That focus—on developing the study's problematic, not objective knowledge construction—is the use to which the narratives were put. The attempt was to get as complete and nuanced an understanding as possible of informants' experiences. The reflexivity of these conversations was a strategy that seemed more useful and appropriate for that goal than a more "objective" approach would have been.

10. "The Independent Living Philosophy promotes and encourages an attitude of self-direction in consumers, enabling Access to community services and resources, facilitating full participation in the community. The IL philosophy recognizes the right of individuals with disabilities to assume risks and make choices. Consumers are encouraged to achieve self-direction over the personal and community services needed to attain their own independent living. To do this, the IL approach develops programs in response to specific needs of individuals." In Canada, the Independent Living movement has defined four general principles to guide Independent Living Resource Centers. They should be :

consumer controlled;
cross disability;
community based;
able to promote integration and full participation (excerpted from Resource Centre for Independent Living web site material).

11. The knowledge base regarding organizational action in texts comes from earlier research by the author and others, using Smith's 1990 work on the conceptual practices of power (See Ng 1996; Diamond 1992; Swift 1995; deMontigny 1995; Walker 1990; Campbell and Manicom 1995).

12. It also seems to reflect a new concept of equity. Culyer (1991) lists four different forms of equity:

1. equal access to treatment;
2. equal treatment received for the same condition;
3. treatment based on the need for care, even if the amount is unequal;
4. equal health status, that is, equal outcomes.

In Justices story, 1 seems to be the criterion used in decision making about medical insurance; 3 seems to be the criterion that Justices would consider fair and reasonable; 4 matches the criterion appearing in provincial health documents.

13. We plan to have teams of three—a staff researcher, a researcher with a disability and a health care provider/researcher—doing the observations. In some cases where it is inappropriate to have three observers at one event, the three team members will observe different instances of the same kind of health care interaction. The goal is to have observers with different perspectives seeing instances of the same health care/client interaction, even if it involves different events.

14. As opposed to describing it simply in the terms that participants know it. Thus, we use a critical analytic framework, developed from our narratives, and the social organization of knowledge (Smith 1990a and 1990b).

REFERENCES

Armstrong, P., and H. Armstrong. 1996. *Wasting Away: The Undermining of Canadian Health Care.* Toronto: Oxford University Press.

Barlow, M., and B. Campbell. 1995. *Straight Through the Heart.* Toronto: Harper Collins Publishing.

British Columbia Provincial Health Officer. 1995. *Health Goals for British Columbia: Identifying Priorities for Healthy Population.* Victoria, BC: Ministry of Health and Ministry Responsible for Seniors.

———. 1997. *A Report on the Health of British Columbians: Provincial Health Officer's Annual Report, 1996.* Victoria, BC: Ministry of Health and Ministry Responsible for Seniors.

British Columbia Auditor General and Deputy Ministers. 1996. *Enhanced Accountability for Performance: A Framework and Implementation Plan, Second Joint Report.* (April) Victoria, BC.

British Columbia Ministry of Health and Ministry Responsible for Seniors. 1993. *New Directions for a Healthy British Columbia.* Victoria, BC.

Campbell, M. 1988. "Management as 'Ruling': A Class Phenomenon in Nursing." *Studies in Political Economy* 27: 29-51.

———. 1992. "Nursing Professionalism in Canada: A Labour Process Analysis." *International Journal of Health Services* 22(4): 475-496.

———. 1998. "Institutional Ethnography and 'Experience' as Data." *Qualitative Sociology* 21(1): 55-72.

Campbell, M., and A. Manicom (eds.). 1995. *Knowledge, Experience and Ruling Relations: Studies in the Social Organization of Knowledge.* Toronto: University of Toronto Press.

Capital Health Region. 1997. *Community Consultation and Communication in the Capital Health Region.* A document approved in principle by the CHR Board, May 7, 1997.

Cerny, P. 1990, *The Changing Architecture of Politics: Structure, Agency and the Future of the State.* London: Sage.

Culyer, T. 1991. "Reforming Health Services: Frameworks for the Swedish Review." In *International Review of the Swedish Health Care System,* edited by A. Culyer. Occasional Paper 34, SNS, Stockholm.

De Montigny, G. 1995. *Social Working: An Ethnography of Frontline Practice.* Toronto: University of Toronto Press.

Diamond, T. 1992. *Making Gray Gold: Narratives of Nursing Home Care.* Chicago: University of Chicago Press.

Evans, B., M. Barer, and T. Marmor. 1995. *Why are Some People Healthy and Others Not?* New York: Aldine de Gruyter.

Hume, S. 1997. "Revisioning Community Development: A Postmodern Analysis of Government Community Development Initiatives from 1988-1993." Unpublished Master of Social Work thesis, University of Victoria, Victoria, BC.

Larson, P. 1997. "Navigating the Implementation of 'New Directions' in a Region of British Columbia: Who's at the Helm? A Study in the Social Organization of Knowledge." Unpublished Master of Nursing thesis. University of Victoria, Victoria, BC.

Lomas, J. 1997. "Devolving Authority for Health Care in Canada's Provinces: 4. Emerging Issues and Prospects." *Canadian Medical Association Journal* 156(6): 817-823.

Lomas, J., J. Woods, and G. Veenstra. 1997. "Devolving Authority for Health Care in Canada's Provinces: 1. An Introduction to the Issues." *Canadian Medical Association Journal* 156(3): 371-377.

Mann, S., and L. Kelley. 1997. "Standing at the Crossroads of Modernist Thought: Collins, Smith, and the New Feminist Epistemologies." *Gender and Society* 11(4): 391-408.

Miller, R. 1997. "Manageable Problems/Unmanageable Death: The Social Organization of Palliative Care." Unpublished Master of Social Work thesis, University of Victoria, Victoria, BC.

Ng, R. 1996. *The Politics of Community Services: Immigrant Women, Class and State.* Halifax, NS: Fernwood.

Saltman, R. 1997. "Equity and Distributive Justice in European Health Care." *International Journal of Health Services* 27 (3): 443-453.

Sibbald, J. 1997. "Delegating Away Patient Safety." *The Canadian Nurse* February: 22-26.

Smith, D. 1987. *The Everyday World as Problematic: A Feminist Sociology.* Toronto: University of Toronto Press.

_____. 1990a. *Conceptual Practices of Power: A Feminist Sociology of Knowledge.* Toronto: University of Toronto Press.

_____. 1990b. *Texts, Facts and Femininity: Exploring the Relations of Ruling.* London: Routledge.

Swift, K. 1995. *Manufacturing Bad Mothers: A Critical Perspective.* Toronto: University of Toronto Press.

Walker, G. 1990. *Family Violence and the Women's Movement.* Toronto: University of Toronto Press

Yeatman, A. 1994. *Postmodern Revisionings of the Political.* London: Routledge.

DISABILITY-RELATED INTENTIONAL INJURY HOSPITALIZATIONS
A MULTI-STATE ANALYSIS

Llewellyn J. Cornelius

ABSTRACT

The increase in intentional injury has led to research on violence prevention as well as research on the effects of violence. The purpose of this paper is to examine factors correlated with disability-related intentional injury hospitalizations (using data from Healthcare Cost and Utilization Project (HCUP-3) for 1992). Findings from this sample of over 800 hospitals across 11 states (California, Colorado, Florida, Iowa, Illinois, Massachusetts, New Jersey, Washington, Arizona, Pennsylvania and Wisconsin) reveal that no one group is immune from the effects of violence. It touches the old, the young, urban residents, rural residents, African Americans, Latinos, Caucasian American, the poor and nonpoor alike. Intentional injury victims with impairment conditions, chronic conditions or degenerative conditions were more likely to be over 65, female, to have Medicaid or self pay as an expected payer for their medical care, to have been hospitalized for child battering or some other form of maltreatment, to have been hospitalized because of being assaulted by a cutting or sharp instrument or to have been hospitalized as a result of the late effects of injuries purposely inflicted by another person.

INTRODUCTION

Statistics on the incidence of violence/intentional injury point to a noticeable national and international trend. Before 1980 homicide was not listed among the 10 leading causes of death in the United States. By 1992 it was the tenth leading cause of death for all Americans (National Center for Health Statistics, 1995a). It was the third leading cause of death for all African American males, and the second leading cause of death for all 15- to 24-year-olds in the United States (NCHS 1995a). This represents a doubling of the homicide rate for all Americans from 1950 to 1992 and a tripling of the rates for those in the 15-24 age group.

The trend in violence is not only reflected in mortality data, it is also found in studies of injuries resulting from violence and in trends on the risk factors associated with violence. For example, several urban trauma centers have reported an increase in cases of intentional injuries (Song, Naude, Gilmore and Bongard 1996; Buss and Abdu 1995; Gladstein, Rusonis and Health 1992; Sege, Stigol, Perry, Goldstein and Spivak 1996; Feero, Hedges, Simmons and Irwin 1995; King 1991). A 1995 national study of college students older than 18 reported that a significant percentage nationwide participated in behaviors that contribute to intentional injury (Douglas et al. 1997). Eight percent carried a weapon. Three percent carried a gun. Ten percent participated in a physical fight and 13 percent had been forced to have sexual intercourse. Another recent study reported that more than half of murders are committed while a person is under the influence of alcohol (Poldrugo 1998). An additional study reported that 16 percent of the homicide victims in the nation's 75 most populous counties were involved in circumstances associated with illegal drugs (ONDCP Drugs and Crimes Clearinghouse 1999).

Several more studies found a disproportionate amount of violence against persons with disabilities, especially among children with multiple disabilities. Children with disabilities are more likely to be physically abused than children without disabilities (Sullivan and Knutson 1998; Crosse, Kaye and Ratnofsky 1993; Sobsey, Randall and Parrila 1997; Goldson 1998). Children with disabilities were also more likely to be physically abused by members of their immediate family (parent, stepparent, sibling or stepsibling) than children without disabilities (Sullivan and Knutson 1998).

Sullivan and Knutson also found that children with multiple disabilities (a combination of behavior disorders, speech/ language disability, mental retardation, hearing impairment, learning disability, health impairment, attention deficit disorder [without a conduct disorder] or some other disability) were more likely to have the most severe forms of physical abuse (a fatality or life-threatening abuse), or had a longer duration of physical abuse than children with a single disability. For some, the consequence of violence was disability (Goldson 1998), while for others exposure to violence was a consequence of their disability (Sullivan and Knutson 1998).

For example, Goldson reports that 18.6 percent of a cohort of 949 children who were abused became permanently disabled following the abuse, while 9.5 percent died following the abuse (Goldson 1998). At the same time, Sullivan and Knutson report that maltreatment was identified as a possible predictor of disability, and that disability was a possible predictor of mistreatment in other cases (Sullivan and Knutson 1998). In addition, both adults and children with disabilities have been abused for longer durations than persons without disabilities (Young, Nosek, Howland, Chanpong and Rintala 1997; Sullivan and Knutson 1998).

Interpersonal violence is now acknowledged as a major cause of injury, disability and death, especially among youth (Powell et al. 1996). Firearm injuries are now the leading cause of death for African American males between the ages of 10 and 34 (Dowd, Knapp and Fitzmaurice 1994). Trauma is now the leading cause of death and disability among children (Adelson and Kochanek 1998). The trend in violence seen in the United States is also occurring in Canada, Guatemala, Mexico, Australia, Norway, Italy, Hong Kong, Ireland, Scotland, Northern Ireland and Great Britain (Eames, Kneafsey and Gordon 1997; Jeanneret and Sand, 1993).

This growth in violence has a widespread impact on society. It is occurring in the home and the workplace. A survey of 500 managers of small and large companies indicated that violence in the workplace has led to an increase in disability compensation claims, an increase in litigation, an increase in security costs and an negative impact on morale and worker productivity (Fenn 1996). The Bureau of Justice Statistics (1996) estimates that violent crimes cost the United States $427 billion annually in medical costs, social welfare program costs and lost earnings. These costs are considered conservative because they do not account for the pain, suffering and loss of quality of life for the victims and their families.

Previous to the 1980s, the issue of violent crimes was seen as a criminal justice issue (Bureau of Justice Statistics 1996). As such, analyses of violent crimes focused on individual behavior. Following the paradigm shift advocated by the National Medical Association (NMA) in 1986 (Schneider, Greenberg and Choi 1992) more recent analyses on the issue have focused what NMA called a "public health approach" to examining violence. This new paradigm emphasized the importance of understanding social and community factors that contribute to the prevalence of violence. The approach requires that one examine individual risk factors (such as self-esteem), social risk factors (such as conflict resolution, poverty, education and peer influence), social norms, societal risk factors (such as the media, public perceptions regarding violence, legislation on violence and abuse) and health care coverage for disability and injury resulting from violence to measure the impact of violence on individuals, communities, and society (Hammond and Yung 1991).

This paradigm shift to community risk factors resulted in an expansion of community-oriented violence prevention efforts, including a federal evaluation of 15 youth violence prevention efforts (Powell et al. 1996). It has also led to an increase in research on disability resulting from violence. However, our ability to understand the risk factors of disability resulting from violence or violence toward persons with disability has been limited by the small sample sizes of the studies. The samples limit the ability to detect findings that would otherwise be statistically significant (Nachmias and Nachmias 1987).

In addition, their focus on a specific clinic population may limit the ability to examine broader social and community factors believed to be related to the risk of violence (income, urban residence, age, gender, etc). This limits the ability to examine some of the geographic and social risk factors to persons with disabilities and disability outcomes resulting from violence. For example, Dowd, Knapp and Fitzmaurice

(1994) have found that 9 percent of a cohort of 72 pediatric cases of intentional injuries resulted in disability. Melzer-Lange and Lye (1996) reported that seven of a cohort of intentional injury cases of adolescent females and nine of the cohort of intentional injury cases of adolescent males seen in a hospital emergency department were for a disability-related condition (head trauma or major trauma). Wright and Litaker (1996) found eight cases of children younger than 14 who were admitted to the hospital for concussion that resulted from intentional injury. Richardson, Davidson and Miller (1996) found that most of a cohort of 12 long-term survivors of an assault rifle attack continued to experience psychological and emotional problems following these incidents. Finally, Zafont et al. (1997) found no difference between 50 subjects with penetrating and nonpenetrating traumatic brain injury in the length of hospital stay, the cost of care, and their degree of functional independence following treatment.

Thus it is expected that studies of samples with larger geographic populations can add some insight to the relationship between violence and disability and disability and violence. The purpose of this study is to examine the patterns of hospitalizations for intentional injury using a large database of hospitalizations across 11 states. This paper has two goals:

1. to determine whether there are differences in the patterns of intentional injury according to socio-demographic factors and selected medical conditions;
2. To examine the differences in access to care for patients with intentional injuries.

DATA AND METHODS

This analysis was based on abstract data from over 5 million discharges from the Healthcare Cost and Utilization Project (HCUP-3) for 1992. HCUP-3 is a project sponsored by the U.S. Department of Health and Human Services (Agency for Health Care Policy and Research 1996). Data for HCUP-3 was obtained from a national probability sample of all discharge records from 20 percent of U.S. community hospitals in 11 States. The states represented in the HCUP-3 data were California, Colorado, Florida, Iowa, Illinois, Massachusetts, New Jersey, Washington, Arizona, Pennsylvania and Wisconsin.

Summarized on each discharge abstract is data related to a patient's hospitalization including age, gender, race, expected source of payment, the length of stay and the total charge for the hospitalization. Hospitals for this study were sampled based on the following characteristics: region (Northeast, South, Midwest, West); location (urban, rural); teaching status; ownership/control (government, nonfederal; private, not-for-profit; private, investor-owned) and bed size (small, medium, large, varying by location and teaching status). Data on selected hospital characteristics (region, location, teaching status, ownership, location, control and bed size) was merged to these hospitalizations from the American Hospital Association annual survey of hospitals. Over 800 hospitals were included in the sample. The sample was designed to represent the universe of short-term, general, nonfederal hospitals in the United States with 30 or more beds.

Accordingly, sampling weights are used in this study to make generalizations to the universe from which this sample was extracted. Measures of intentional injury and impairment conditions, chronic conditions, and degenerative conditions (to be described below) were used to form the analytical frame of this study.

Measures of Impairment Conditions, Chronic Conditions, Degenerative Conditions, Intentional Injury, Health Care Use and Expenses

Impairment Conditions, Chronic Conditions and Degenerative Conditions

As part of the data abstraction process, the condition for each hospitalization was coded using the International Classification of Disease, nine edition (ICD-9). Up to 15 conditions were recorded for each hospitalization. An algorithm of impairment conditions, chronic conditions and degenerative medical conditions was used with these conditions to compute a biophysical measure of chronic illness and impairment (see Altman [1993] for a detailed description of the development of this measure).

The ICD-9 conditions were sorted into chronic or degenerative conditions, because those conditions are associated with potential functional or role limitations. It should be noted that while these conditions are related to disability, not all cases result in a disability. Listed below is a summary of this indicator. Chronic illness was defined as taking two forms. Most chronic conditions are those with gradual onset, which result in permanent involvement with pathology—with or without acute episode—with the potential to be controlled, but not cured. Examples of these chronic conditions include arthritis, kidney disease and diabetes. A second form of chronic conditions are those which have acute onset, such as heart attack and stroke, which generally result in some permanent damage which can be controlled or rehabilitated but not totally cured. The construction of this chronic condition variable was done with the consultation of a physician, and included only conditions identified by the ICD-9 code known to fulfill the definitions. Condition codes which indicated an unspecified category within an organ system, such as 573.9, an unspecified disorder of liver, were not specific enough to classify and were therefore included in the "other" category.

The second condition, impairment, was defined as a residual change in the condition's makeup, resulting from an accident, disease or birth defect, without an ongoing pathology. It included such conditions as cerebral palsy, spinal cord injury and amputations. Once again, ICD-9 codes were used to identify the specific instance of impairment. As with the chronic condition variable, the ICD-9 codes without sufficient detail such as 746.9, an unspecified anomaly of the heart, were coded as "other."

Using these two variables, a combination measure was created identifying those with a chronic condition or impairment only, and those who had both a chronic condition and an impairment. The resulting measure has four categories:

1. chronic conditions;
2. impairment;
3. both chronic conditions and impairment;
4. all others, which can include acute conditions and the unspecified conditions.

The phrase "disability-related" is used throughout the paper to refer to hospitalizations for those with chronic conditions or degenerative conditions or impairment conditions.

Intentional Injury

Indicators of intentional injury were formed using the International Classification of Disease, nine edition (ICD-9). Each of the 15 conditions reported during each hospitalization were compared against the following intentional injury E-Codes:

E-960 - Fight brawl rape, homicidal;
E-961 - Assault by corrosive or caustic substance except poisoning;
E-962 - Assault by poisoning;
E-963 - Assault by hanging and strangulation;
E-964 - Assault by submersion (drowning);
E-965 - Assault by firearms and explosives;
E-966 - Assault by cutting and piercing instrument;
E-967 - Child battering and other maltreatment;
E-968 - Assault by other and unspecified means;
E-969 - Late effects of injury purposely inflicted by other person.

An indicator was computed for each hospitalization that had at least one condition that matched the relevant E code listed above. In no case was an intentional injury condition listed as the primary diagnosis for the hospitalization. An intentional injury condition was listed as the secondary condition for one-third of the hospitalizations, while it was listed as the third condition for 20 percent of the hospitalizations, and as the fourth condition for 10 percent of the hospitalizations. It should be noted, however, that the sequence of the conditions reported on the hospitalization discharge does not indicate which condition takes precedence over the other in the treatment process.

Use of and Expenses for Services

Measures of the use of services and total charges for services were based on the data supplied in the hospital discharge record. The mean length of stay (LOS) was calculated by dividing the sum of inpatient days by the number of patients. Inpatient days are calculated by subtracting day of admission from day of discharge, so persons entering and leaving a hospital on the same day have a length of stay of zero. The mean total charge was calculated by dividing the sum of patient charges by the number of patients. Total charges only represent the dollar amount charged for the hospitalization not the amount paid or the actual costs to provide the care.

Tests of Significance

Tests of significance were conducted using SUDAAN (Shah et al. 1992) to determine the statistical significance of the findings presented in this paper. Two tests of statistical significance were used to detect whether the data reported in these analyses were statistically significant. The first was the standard error of a percent. A major purpose of HCUP-3 is to allow for the construction of population estimates based on sample data. Like all probability samples, there is a margin of error between the response given by a sampled respondent and what the actual response to a question would be if a census was taken. The "standard error" represents the difference between the reported results and what the results would have been if a census of the total population was taken. Percentages displayed in tables and figures with a relative standard error of more than 30% are noted in the tables with an asterisk. This indicates that the actual response by the population for a given question may be at least 30% higher or lower than what is listed in the table.

The second statistic used to perform tests of statistical significance in this study is the Student T test. The Student T test was used to determine the statistical significance of two percentages or means being compared in the analysis. Unless otherwise noted, only statistically significant differences ($p<.05$) are discussed in the text.

RESULTS

Overall Patterns of Disability-Related Intentional Injury

In 1992, intentional injuries and disability-related intentional injuries represented a small subset of all hospitalizations. At the same time though, there were significant differences between victims of intentional injury with impairment conditions, with chronic conditions or degenerative conditions and other victims of intentional injury. In 1992 there were 43,220 intentional injury discharges (table 1) in the 11 states included in this study (California, Colorado, Florida, Iowa, Illinois, Massachusetts, New Jersey, Washington, Arizona, Pennsylvania and Wisconsin). This accounted for less than 1 percent of the 35.1 million discharges during that year (Agency for Health Care Policy and

Research 1999). Slightly over 12 percent (12.3) of the intentional injury hospitalizations were disability related.

Hospitalizations as a result of attempted homicide or rape, as a result of firearm assault or assault from a sharp instrument were the most frequently occurring intentional injury hospitalizations. However, hospitalizations for either child battering or some other form of maltreatment and for the late effect of injuries purposely inflicted by other persons were disproportionately represented among disability-related hospitalizations.

Among the intentional injury victims, persons with impairment conditions were more likely than others to be victims of assault by a cutting or piercing instrument (46.5 percent versus 25.8 percent and 38.1 percent). However, they were less likely than intentional injury victims

Table 1. Percent of Discharges for Intentional Injury by Type of Condition, 1992

	Type of condition			
	Chronic degenerative only	Impairment only	Both chronic/ impairment	All other conditions
Number of hospitalizations	2,746	2,192	185	38,097
E960-Fight brawl rape - homicidal	47.2%	28.8%	44.8%	42.1%
E961-Assault by corrosive or caustic substance except poisoning	0.2%*	0.0%	0.0%	0.1%
E962-Assault by poisoning	1.1%	0.0%	0.0%	0.3%
E963-Assault by hanging and strangulation	0.4%*	0.0%	0.0%	0.2%
E964-Assault by submersion (drowning)	0.0%	0.0%	0.0%	0.0%
E965-Assault by firearms and explosives	7.8%	11.8%	6.0%	13.7%
E966-Assault by cutting and piercing instrument	25.8%	46.5%	10.4%	38.1%
E967-Child battering and other maltreatment	12.6%	3.7%	23.7%	3.8%
E968-Assault by other and unspecified means	0.7%	0.0%	0.0%	0.3%
E969-Late effects of injury purposely inflicted by other person	4.4%	9.1%	15.2%	1.6%

Note: * = standard error ≥ 30 percent.
Source: Agency for Health Care Policy and Research: Healthcare Cost and Utilization Project (HCUP-3), 1992.

with a chronic condition or degenerative condition to be victims of attempted homicide or rape (28.8 percent versus 47.2 percent). Intentional injury victims with impairment conditions, chronic conditions or degenerative conditions were more likely than others to be victims of the late effects of injuries purposely inflicted by another person (4.4 percent, 9.1 percent and 15.2 percent versus 1.6 percent respectively). Finally, intentional injury victims with a chronic or degenerative condition were more likely than others to be the victim of child battering or some other form of maltreatment (12.6 percent versus 3.7 percent and 3.8 percent).

Patient Characteristics of Intentional Injury Hospitalizations

Data on the characteristics of patients hospitalized for intentional injury highlight the complex picture of violence in America. On one hand, it is disproportionately seen among the poor, the young, African Americans and Latinos. On the other hand, it also touches the nonpoor, the elderly and white Americans (table 2). Children and young adults (under age 35) accounted for 39.5 percent of all the hospitalizations in this sample. However, they accounted for 48.2 percent of the intentional injury victims with chronic degenerative conditions and 71.3 percent of the intentional injury victims with impairment conditions (table 2). African Americans and Latinos accounted for 13.4 percent of all the hospitalizations in 1992. Yet they accounted for over 44 percent of all intentional injury hospitalizations (44.7 percent, 53.3 percent, and 55.2 percent respectively). Patients residing in lower-income communities (patients who lived in zip codes where the median income was under $20,000) accounted for 10.2 percent of all hospitalizations. However, they accounted for close to 30 percent of the hospitalizations for intentional injury (30.6 percent, 28.7 percent, and 28.9 percent respectively). It should be noted that hospitalizations for intentional injury were also found among the elderly, whites and persons living in zip codes with a median income greater than $45,000.

In addition to highlighting the complex profile of violence in America, this data also highlights the disproportionate representation of disability-related intentional injury hospitalizations among the elderly and females (table 2). Persons over 65 represented only 1.3 percent of the nondisability-related hospitalizations for intentional injury. At the same time, they represented 9.7 percent of the persons with chronic degener-

Table 2. Selective Patient Characteristics of Discharges for Patients with Intentional Injuries, 1992

	Type of Condition for Those with Intentional Injuries			
	Chronic degenerative only	Impairment only	Other conditions	All Hospitalizations
Number of hospitalizations	2,746	2,192	38,097	35,011,737
Age of patient				
0-17	17.0%	12.4%	13.7%	19.2%
18-34	31.2%	58.9%	58.8%	20.3%
35-44	21.6%	18.3%	18.6%	9.2%
45-64	20.5%	9.5%	7.6%	17.4%
61 and over	9.7%	.8%	1.3%	33.8%
Gender of patient				
Male	68.1%	84.3%	81.7%	41.9%
Female	31.9%	15.7%	18.3%	58.1%
Race of patient[1]				
White	51.4%	44.5%	41.4%	52.2%
Black	34.3%	34.9%	32.0%	7.1%
Hispanic	10.4%	18.4%	23.2%	6.3%
Asian/Pacific Islander	1.3%	1.3%	1.5%	1.3%
Native American	1.1%	.7%	.4%*	0.2%
Other	1.5%	.3%	1.5%	0.9%
Expected Payor				
Medicare	10.6%	3.4%	3.1%	34.4%
Medicaid	34.8%	26.4%	25.6%	16.8%
Private Insurance/HMO	22.6%	20.2%	23.6%	39.0%
Self-pay	22.5%	33.5%	30.8%	5.5%
No Charge	0.5%	0.2%	0.5%	0.1%
Other	9.0%	16.3%	16.4%	3.8%
Median income of patients Zip code (community)				
$0-15000	10.5%	8.4%	8.7%	2.6%
$15001-20000	20.1%	20.3%	20.2%	7.5%
$20001-25000	22.4%	26.0%	26.4%	18.4%
$25001-30000	19.2%	16.2%	18.1%	19.1%
$30001-35000	15.5%	14.2%	12.3%	15.9%
$35001-40000	5.8%	8.9%	7.0%	10.7%
$40001-45000	3.8%	1.6%	3.3%	7.0%
$45001+	2.7%	4.3%	4.0%	9.8%

Notes: [1]Excludes hospitals that did not provide any data on race/ethnicity (30 percent of hospitalizations).
Excluded from this table are hospitalizations for intentional injury for persons with both chronic disease and impairment conditions (due to small sample size).
*=standard error ≥ 30 percent.
Source: Agency for Health Care Policy and Research: Healthcare Cost and Utilization Project (HCUP-3), 1992.

Table 3. Selective Hospital Characteristics of Discharges for Persons with Intentional Injuries, 1992

	Type of Condition for Those with Intentional Injuries			
	Chronic degenerative only	Impairment only	Other conditions	All Hospitalizations
Number of Hospitalizations	2,746	2,192	38,097	35,011,737
Region				
Northeast	12.9%	22.6%	17.6%	19.9%
Northcentral	22.7%	12.7%	15.1%	34.4%
South	37.6%	28.6%	26.8%	22.8%
West	26.8%	36.1%	40.6%	23.0%
Type of hospital				
Rural	9.2%	7.1%	8.2%	35.1%
Urban nonteaching	40.3%	33.8%	34.7%	51.1%
Urban teaching	50.5%	59.1%	57.1%	12.8%
Hospital bedsize				
Small	13.9%	12.4%	13.5%	37.0%
Medium	37.3%	33.2%	38.6%	31.6%
Large	48.8%	54.4%	47.9%	30.5%
Type of hospital ownership				
Public	28.0%	31.2%	34.6%	18.7%
Private nonprofit	66.7%	65.1%	61.7%	67.9%
Private for profit	5.3%	3.6%	3.6%	13.5%

Note: Excluded from this table are hospitalizations for intentional injury for persons with both chronic disease and impairment conditions (due to small sample size).
Source: Agency for Health Care Policy and Research: Healthcare Cost and Utilization Project (HCUP-3), 1992.

ative conditions who were hospitalized for intentional injury. Similarly, women presented 18.3 percent of the nondisability-related hospitalizations for intentional injuries and 31.9 percent of the persons with chronic degenerative conditions who were hospitalized for intentional injury.

Hospital Characteristics of Intentional Injury Hospitalizations

Like patient characteristics, hospital characteristics are useful in highlighting the complexities of intentional injury in America. Hospitalizations for intentional injury were not only found in urban hospitals and in public hospitals, but also in for-profit hospitals, small hospitals and rural hospitals. Nearly two-thirds (63.9 percent) of all the hospitalizations in this study were in facilities classified by the American Hospital Association as urban hospitals. These hospitals accounted for more than

Table 4. Hospital Utilization and Expenditures by Type of Condition for Persons with Intentional Injuries, 1992

	Type of Condition for Those with Intentional Injuries			
	Chronic degenerative only	Impairment only	Other conditions	All Hospitalizations
Number of hospitalizations	2,746	2,192	38,097	35,011,737
Admissions type				
Emergency	69.6%	63.6%	69.7%	30.2%
Urgent	26.4%	26.7%	25.8%	27.4%
Elective	4.0%	9.7%	4.5%	21.2%
Newborn	0.0%	0.0%	0.0%	10.2%
Delivery	0.0%	0.0%	0.0%	2.4%
Disposition at discharge				
Routine	84.7%	90.4%	90.8%	81.5%
Short-term hospital	1.6%	1.6%	1.3%	2.2%
Skilled nursing facility	3.3%	.9%	.2%	0.9%
Intermediate care facility	.5%*	.3%*	.2%*	2.1%
Other Facility	2.7%	1.7%	3.0%	0.9%
Home health care	2.4%	2.0%	1.2%	2.7%
Against medical advice	2.1%	3.1%	2.5%	0.1%
Died in hospital	2.7%	.0%	.7%	0.1%
Mean length of stay	5.9	4.2	3.9	5.9
Mean total charge	$12,906	$10,286	$9,646	$9,630

Notes: * = standard error ≥ 30 percent.
Excluded from this table are hospitalizations for intentional injury for persons with both chronic disease and impairment conditions (due to small sample size).
Source: Agency for Health Care Policy and Research: Healthcare Cost and Utilization Project (HCUP-3), 1992.

78 percent (table 3) of the hospitalizations for intentional injury (90.5 percent, 92.9 percent, 78.0 percent and 91.8 percent respectively). Public hospitals accounted for 18.7 percent of all the hospitalizations in this study. Yet they accounted for at least one-quarter of the hospitalizations for intentional injury (28.0 percent, 31.2 percent, and 34.6 percent respectively). Slightly over 30 percent of all hospitalizations were in large hospitals, yet they accounted for at least 47 percent of the hospitalizations for intentional injury (48.8 percent, 54.4 percent, and 47.9 percent respectively). Finally, a significant portion of intentional injury was also found in rural hospitals, small hospitals and private hospitals.

Patterns of Utilization and Expenses for Intentional Injury

To examine the issue of disability-related intentional injury it is important not only to examine the profile of the hospitalizations, but also the patterns of access to care and outcomes of hospitalization for patients who were hospitalized for intentional injury. Data in tables 4-6 examine the patterns of utilization and expenses for these types of hospitalizations. As indicated in table 4, regardless of medical condition, victims of intentional injury were more likely than others to be admitted to the hospital for either an emergency or for urgent care.

At the same time, none of the intentional injury hospitalizations were for either the delivery of a child or for the care of a newborn baby. Patients who were admitted for intentional injury were just as likely as other patients to have a routine medical discharge (84.7 percent, 90.4 percent, 90.8 percent and 81.5 percent respectively). However, intentional injury victims were more likely than persons without intentional injuries to either leave the hospital against medical advice or to die in the hospital. At the same time, victims of intentional injuries with chronic degenerative conditions were more likely than persons without intentional injuries to die in the hospital (2.7 percent versus 0.1 percent). Intentional injury victims with impairment conditions had a shorter length of stay (table 4) than the average patient (4.2 days versus 5.9 days). Persons with chronic degenerative conditions who were hospitalized for assault by firearms and explosives had longer lengths of stays than others (8.2 days versus 4.9 and 4.7 days [table 5]). Persons with impairment conditions, chronic conditions or degenerative conditions who were hospitalized for the late effect of injury purposely inflicted by other persons also had longer lengths of stays than others (10.9 days and 10.2 days versus 7.1 days respectively [table 5]).

Intentional injury patients with disabilities had higher mean total charges than others ($12,906 and $10,286 versus $9,646 and $9,630 [table 4]). The highest mean total charge was for persons hospitalized because of the late effects of injury purposely inflicted by other persons, because of assault by firearms or because of assault from a cutting or piercing instrument (table 6). These patterns were correlated with the length of stay (table 5). Persons with longer lengths of stays also had higher mean expenses than others. However there was one exception worthy of note. The highest mean total charge was for persons with chronic conditions who were hospitalized for being assaulted by firearms ($21,502) yet they had a lower length of stay than persons with chronic conditions who were hospitalized because of the late effects of injuries purposely inflicted by another person (8.2 days versus 10.9 days [tables 6 and 5]).

Table 5. Mean Length of Stay for Selective Intentional Injury Discharges by Type of Condition, 1992

	Type of Condition for Those with Intentional Injuries		
	Chronic degenerative only	Impairment only	All other conditions
Number of hospitalizations	2,746	2,192	38,097
E960-Fight brawl rape - homicidal	4.8	3.7	3.5
E965-Assault by firearms and explosives	8.2	4.9	4.7
E966-Assault by cutting and piercing instrument	5.9	3.0	3.6
E967-Child battering and other maltreatment	6.8	5.4	5.4
E969-Late effects of injury purposely inflicted by other person	10.9	10.2	7.1

Note: Excluded from this table are hospitalizations for intentional injury for persons with both chronic disease and impairment conditions (due to small sample size).
Source: Agency for Health Care Policy and Research: Healthcare Cost and Utilization Project (HCUP-3), 1992.

Table 6. Total Charges for Selective Intentional Injury Discharges by Type of Condition, 1992

	Type of Condition for Those with Intentional Injuries		
	Chronic degenerative only	Impairment only	All other conditions
	Mean total charge		
Number of hospitalizations	2,746	2,192	38,097
E960-Fight brawl rape - homicidal	$10,088	$9,042	$7,971
E965-Assault by firearms and explosives	$21,502	$13,400	$13,207
E966-Assault by cutting and piercing instrument	$15,464	$9.615	$10,276
E967-Child battering and other maltreatment	$8,914	$4,390	$5,719
E969-Late effects of injury purposely inflicted by other person	$19,798	$16,365	$11,800
Total charges (in thousands)	$34,603	$21,756	$416,040

Note: Excluded from this table are hospitalizations for intentional injury for persons with both chronic disease and impairment conditions (due to small sample size).
Source: Agency for Health Care Policy and Research: Healthcare Cost and Utilization Project (HCUP-3), 1992.

DISCUSSION AND IMPLICATIONS

Data from a sample of 800 hospitals across 11 states (California, Colorado, Florida, Iowa, Illinois, Massachusetts, New Jersey, Washington, Arizona, Pennsylvania and Wisconsin) reveal a multitude of issues regarding both the prevalence of intentional injury as well as the prevalence of disability-related intentional injury.

First, hospitalizations for intentional injury came to over $440 million (and over $65 million for persons with disability-related hospitalizations) in 1992. These are clearly costs that could be prevented with appropriate violence-prevention strategies.

Second, Medicaid and "self-pay" were designated at the source of payment for a significant portion of these charges. Urban as well as public teaching hospitals provide a considerable portion of the care delivered to persons who were hospitalized for intentional injury. These hospitals are considered by some to be safety net hospitals because of the large portion of care they provide to the indigent. Urban public hospitals account for one-third of the uncompensated care in the United States. Rural public hospitals and teaching hospitals also provide a significant portion of uncompensated care. These hospitals also provide a considerable amount of care to Medicaid patients (Mann, Melnick, Bamezai and Zwaniger 1997). All these factors suggest that urban and teaching hospitals play a critical role in providing care to victims of intentional injury who would otherwise fall between the cracks.

Third, while a large portion of victims of violence were young, poor, African American, Latinos or patients seen in urban hospitals, this did not mean that the elderly, whites, the economically well off or rural residents were not also touched by violence.

Finally, and most importantly to this analysis, there were clusters of intentional injuries found among persons with disability-related conditions who were hospitalized in 1994. Persons with disabilities were more likely than others to be hospitalized because of child battering, the late effects of injuries purposely inflicted by another person or some other form of maltreatment. In addition, persons with impairments were also more likely than others to be victims of assault by a cutting or piercing object. This supports the hypothesis posed by Sullivan and Knutson that persons with disabilities may likely be victims of abuse as well as an indication that persons may become disabled following a violent act.

What are the implications of these findings? Data from the study suggests that intentional injury and disability-related intentional injury are significant problems. Other studies of intentional injury suggest that while all ages, genders and ethnic groups are victims of violence, the potential for violence continues to be a problem in urban areas, among African and Latino Americans, among young adults and among males. While the overall national rates have declined slightly in the last few years, in 1997 African Americans were seven times more likely than whites to be murdered (Fox and Zawitz 1999). Men were nine times more likely than females to be murdered. Young adults (ages 18-24) had the highest victimization rates for homicides. National data on death from firearm-related injuries indicate that in 1996 Native American males, African American males and Latino males were more likely and to die from firearm-related injuries than Asian American or white males (National Center for Health Statistics 1998).

Several studies of hospitals in urban areas have found that some of the persons recently hospitalized for intentional injury were repeat victims of violence. One study of recurrent intentional injury in south central Los Angeles found that 22 percent of the cases of intentional injury had been admitted to the hospital previously for intentional injury (Kennedy, Brown, Brown and Fleming 1996). A second urban area study (San Francisco) reported a recidivism rate of 16 percent for a cohort of patients under 25 years of age (Tellez, Mackersie, Morabito, Shagoury and Heye 1995). Gunshot wounds represented 44 percent of these injuries. At the same time, 92 percent of the deaths that occurred as a result of violence were from firearms. A study of a third urban area (Philadelphia) reported a recidivism rate of 40.9 percent for a cohort of men between 20 and 29 years of age (Schwarz, Grisso, Miles, Holmes, Wishner and Sutton, 1994).

These findings suggest that even if the there is a national decline in reported rates of violence, it is still a pressing problem for segments of the population—residents in urban areas, young adults, males, African Americans and Latinos, and from this analysis, persons with disability-related conditions. At the same time, data on the magnitude of disability resulting from intentional injury is still scant. Current studies do not examine the potential for functional or role limitations as a result of intentional injury. Neither have they examined the psychosocial consequences of disability resulting from intentional injury. Some of the consequences include: feeling angry or depressed because of the change in lifestyles after the injury and lacking supportive mechanisms for helping to adjust to the changes in lifestyle. Finally, studies also have not examined all the possible causes of violence, such as the relationship between caregiver stress and burden and the potential for inflicting violence against persons with disabilities. While there is a need for more work in this area, what is clear is that disability resulting from violence is preventable and even one case of disability resulting from violence is one case too many.

In closing, this study suggests that there is an underlying two-way relationship between violence and disability. Some specific types of intentional injury were found to be correlated with disability (assault from a cutting or piercing instrument or the late effects of injury purposely inflicted by another person). Other forms of injury (maltreatment or child battering) may occur because someone has a disability. However, the lack of data in this study on the onset of the disability condition relative to the intentional injury condition for each hospitalization limits the ability to establish causality between intentional injury and disability. However, others have indicated that disability is one of the outcomes of interpersonal violence (Powell et al. 1996).

REFERENCES

Adelson, P.D., and P.M. Kochanek. 1998. "Head Injury in Children." *Journal of Child Neurology* 13: 2-15.

Agency for Health Care Policy and Research. 1996. "Summary of Documentation for the HCUP-3 Nationwide Inpatient Sample (NIS), Release 1, 1988-1992." Rockville, MD.

_____. 1999. "National Statistics by Principal Procedure: HCUP-3 Nationwide Inpatient Sample for 1992 Hospital Inpatient Stays." Available: http://www.ahcpr.gov/data/hcup/pcchptb.htm.

Altman, B.M. 1993. "Definitions of Disability and Their Measurement and Operationalization in Survey Data." In *Proceedings of the 1993 Public Health Conference on Records and Statistics: Toward the Year 2000, Refining the Measures*. Washington, DC: U.S. Department of Health and Human Services.

Bureau of Justice Statistics. U.S. Department of Justice. 1996. *Sourcebook of Criminal Justice Statistics, 1995*. Washington, DC: Government Printing Office.

Buss, T.F., and R. Abdu. 1995. "Repeat Victims of Violence in an Urban Trauma Center." *Violence and Victims* 10: 183-194.

Crosse S.C., E. Kaye, and A.C. Ratnofsky. 1993. *A Report on the Maltreatment of Children with Disabilities*. Washington, DC: National Center on Child Abuse and Neglect.

Douglas, K.A. 1997. "Results from the 1995 National College Health Risk Behavior Survey." *College Health* 46: 55-66.
Dowd, M.D., L.F Knapp, and L.S. Fitzmaurice. 1994. "Pediatric Firearm Injuries, Kansas City, 1992: A Population Based Study." *Pediatrics* 94: 867-873.
Eames, M.H., B. Kneafsey, and D. Gordon. 1997. "A Fractured Peace: A Changing Pattern of Violence." *British Journal of Plastic Surgery* 50: 416-420.
Feero, S., J.R. Hedges, E. Simmons, and L. Irwin. 1995. "Intracity Regional Demographics of Major Trauma." *Annals of Emergency Medicine* 25: 788-793.
Fenn, D. 1996. "The Effects of Workplace Violence." *Inc* 18: 116.
Fox, J.A., and M.W. Zawitz. 1999. *Homicide Trends in the United States.* Bureau of Justice Statistics. Available: Http://www.ojp.usdoj.gov/bjs/homicide/homtrnd.htm.
Gladstein, J., E.J. Rusonis, and F.P. Health. 1992. "A Comparison of Inner-city and Upper-middle-class Youths' Exposure to Violence." *Journal of Adolescent Medicine* 13: 275- 280.
Goldson, E. 1998. "Children with Disabilities and Child Maltreatment." *Child Abuse and Neglect* 22: 663-667.
Hammond, W.R., and B.R. Yung. 1991. "Preventing Violence in At-risk African-American Youth." *Journal of Health Care for the Poor and Underserved* 2: 359-373.
Jeanneret, O., and E.A. Sand. 1993. "Intentional Violence Among Adolescents and Young Adults: An Epidemiological Perspective." *World Health Statistics Quarterly* 46: 34-50.
Kennedy, F., J.R. Brown, K.A. Brown, and A.W. Fleming. 1996. "Geographic and Temporal Patterns of Recurrent Intentional Injury in South-central Los Angeles." *Journal of the National Medical Association* 88(9): 570-572.
King, W.D. 1991. "Pediatric Injury Surveillance: Use of a Hospital Discharge Data Base." *Southern Medical Journal* 84: 342-348.
Mann, J.M., G.A. Melnick, A. Bamezai, and J. Zwaniger. 1997. "A Profile of Uncompensated Care, 1983-1995." *Health Affairs* 16(4): 223-232.
Melzer-Lange, M., and P.S. Lye. 1996. "Adolescent Health in a Pediatric Emergency Department." *Annals of Emergency Medicine* 27: 633-637.
Nachmias, D., and C. Nachmias. 1987. *Research Methods in the Social Sciences* (3rd edition). New York: St. Martin's Press.
National Center for Health Statistics. 1995a. *Health United States, 1994.* Hyattsville, MD: U.S. Department of Health and Human Services, Centers for Disease Control and Prevention.
_____. 1998. *Health United States, 1998.* Hyattsville, MD: U.S. Department of Health and Human Services, Centers for Disease Control and Prevention.
ONDCP Drugs and Crime Clearinghouse 1994. *Drugs and Crime Facts, 1994.* Rockville, MD: National Criminal Justice Reference Service.
Poldrugo, F. 1998. "Alcohol and Criminal Behaviour." *Alcohol and Alcoholism* 33(1): 12.
Powell, K.E., L.L. Dalhberg, J. Friday, J.A. Mercy, T. Thornton, and S. Crawford. 1996. "Prevention of Youth Violence: Rationale and Characteristics of 15 Evaluation Projects." *American Journal of Preventive Medicine* 12: 3-12.

Richardson, J.D., D. Davidson, and F.B. Miller. 1996. "After the Shooting Stops: Follow Up on Victims of an Assault Rifle Attack." *Journal of Trauma* 41: 789-793.
Schwarz, D.F., J.A. Grisso, C.G. Miles, J.H. Holmes, and A.R. Wishner. 1994. "A Longitudinal Study of Injury Morbidity in an African-American Population." *Journal of the American Medical Association* 271(10): 755-760.
Schneider, D., M.R. Greenberg, and D. Choi. 1992. "Violence as a Public Health Priority for Black Americans." *Journal of the National Medical Association* 84: 843-848.
Sege, R., L.C. Stigol, C. Perry, R. Goldstein, and H. Spivak. 1996. "Intentional Injury Surveillance in a Primary Care Pediatric Setting." *Archives of Pediatric Adolescent Medicine* 150: 277-283.
Shah, B.V., B.G. Barnwell, P.N. Hunt, amd L.M. LaVange. 1992. *SUDAAN Users Manual: Professional Software for Survey Data Analysis for Multi-stage Designs*. Research Triangle Park, NC: Research Triangle Institute, 1992.
Sobsey, D., W. Randall, and R.K. Parrila. 1997. "Gender Differences Among Abused Children with and without Disabilities." *Child Abuse and Neglect* 21: 707-720.
Song, D.H., G.P. Naude, D.A. Gilmore, and F. Bongard. 1996. "Gang Warfare: The Medical Repercussions." *Journal of Trauma* 40: 810-815.
Sullivan, P.M., and J.F. Knutson. 1998. "The Association Between Child Maltreatment and Disabilities in a Hospital Based Epidemiological Study." *Child Abuse and Neglect* 22: 271-288.
Tellez, M.G., R.C. Mackersie, D. Morabito, C. Shagoury, and C. Heye. 1995. "Risks, Costs, and the Expected Complications of Re-injury." *American Journal of Surgery* 170(6): 660-663.
Wright, M.S., and D. Litaker. 1996. "Childhood Victims of Violence: Hospitalization by Children with Intentional Injuries." *Archives of Pediatric Adolescent Medicine* 150: 415- 420.
Young, M.E., M.A. Nosek, C. Howland, G. Chanpong, and D.H. Rintala. 1997. "Prevalence of Abuse of Women with Physical Disabilities." *Archives of Physical Medicine and Rehabilitation* 78(Suppl 5): S34-38.
Zafonte, R.D. et al. 1997. "Functional Outcome After Violence Related Intentional Injury." *Brain Injury* 11: 403-407.

DISENFRANCHISED
PEOPLE WITH DISABILITIES IN AMERICAN ELECTORAL POLITICS

Todd G. Shields, Kay Schriner, Ken Schriner, and Lisa Ochs

ABSTRACT

People with disabilities have been called "the sleeping giant in our midst" (Zola 1993) because of their vast numbers and relative isolation from the American mainstream. Usually, discussions of this isolation are framed in terms of higher rates of unemployment, lower levels of educational attainment, lower incomes and lower rates of community participation (Louis Harris & Associates 1986, 1994). However, this characterization of people with disabilities as a huge, unrealized force may be especially apt in the context of electoral politics. In this paper, we review some of our recent research that has begun to shape our understanding of the role of people with disabilities in the American political system. This review of our prior work includes empirical tests of the "conventional wisdom" regarding the factors that influence electoral participation as well as a discussion that reflects recent theoretical developments in disability studies and political science. While in this review we do not provide an exhaustive summary of our prior work, we do provide an overview of this long-term research agenda. Finally, we conclude with many unanswered questions and ideas for future research.

THE IMPORTANCE OF POLITICAL PARTICIPATION

Citizen participation in governance is vital for a healthy democratic society (Verba, Schlozman and Brady, 1995). Alienation from political life reflects directly on the degree and authenticity of representation in a democracy (Pitkin 1967; Verba et al. 1995). If democratic governments are to be responsive to the concerns of individual citizens, those citizens must make their interests known through participation in the electoral process. As Sidney Verba argues:

> [I]f the government is to have the capability of giving equal consideration to the needs and preferences of all citizens, the public must be equally capable of providing that information. They must provide information about themselves—who they are, what they want, what they need. If citizen activity is the main way in which that is done, then democratic responsiveness depends on equal participation (Verba 1996, p. 2).

The importance of electoral participation is underscored by political theorists who emphasize the relationship between participation and the content of public policy. In arguing that public policy should be the focus of studies of democratic systems of government, Mayo notes that it is "the reference to policies—their complex formation and execution—that gives content to the governing function of a society" and that "at the heart of the governing function ... is the policies which are reached and carried out" (Mayo 1960, p. 4).

The vital connection between electoral participation and the policies formulated by elected representatives heightens concern about the role of people with disabilities in electoral politics because so many of these individuals are so greatly affected by public policy. One need only examine the level of public spending on disability programs to recognize the significance of disability in domestic policy, and the importance of public policy to people with disabilities (Schriner and Batavia 1995).

For example, some 7 million people with disabilities receive income maintenance and health care services that cost in excess of $100 billion per year (General Accounting Office 1996). A smaller, though hardly insignificant, amount funds a variety of residential, rehabilitation and treatment programs for many more individuals with disabilities. Civil rights statutes such as the Individuals with Disabilities Education Act and the Americans with Disabilities Act protect the rights of persons with disabilities to participate in and benefit from societal institutions.

The need to ensure authentic representation and the direct relationship between political participation and the content of public policy are more than sufficient reason to inquire about the role of people with disabilities in the nation's electorate. Justifying this inquiry in this way also calls attention to how little we actually know about the electoral participation of citizens with disabilities. While scholars have examined the role disability rights activists have played in passing federal legislation (e.g., Percy 1989; Scotch 1984; Watson 1993), and many researchers have analyzed the value basis of disability policy (e.g., Hahn 1985; Schriner, Rumrill and Parlin 1995) or analyzed the effects of federal policy on the lives of people with disabilities (e.g., Braddock and Hemp 1996; Nosek, Fuhrer, Rintala and Hart 1993), there has been almost no attention paid to the simplest questions about their political behavior.

Do people with disabilities register and vote at the same rates as people without disabilities? Are people with disabilities more likely to vote absentee? If so, do they view absentee voting as a reasonable way of accommodating them or as a failure to make up for inaccessible polling places? Do rates and patterns of participation vary by type, severity or onset of disability? What legal barriers exist to the participation of people who are labeled incompetent, or who are under guardianship? If legal barriers exist, are they constitutional? The myriad of unanswered questions about political participation of people with disabilities obviously cannot be answered in this review. However, we can set the stage for future research by describing what we do know and situating this discussion in the context of other research on political participation more generally and on disability politics and policy.

DEFINING DISABILITY

A necessary first step in understanding the behavior of persons with disabilities is to define the population of individuals we refer to when we use the term.[1] Unfortunately, this task is not easy.[2] Complicating the problem is the tendency of different investigations to use different measures of disability. Such is often the case when looking at the few investigations of political behavior and disability.

For example, the two data sets that we have relied upon heavily each employ a different measure of disability—making comparisons between the data sets (and across years) extremely difficult. In the 1987 Lou Harris Survey of the political attitudes and behaviors of people with disabilities, the investigators defined disability in three ways. Respondents were identified as disabled if:

1. they reported having a limiting health condition which interferes with their normal activities, a condition which prevents or limits their ability to work or a physical disability such as a seeing, hearing or speech impairment, an emotional or mental disability or a learning disability;
2. the participant said they considered themselves disabled;
3. they said other people would consider them disabled.

The survey also attempted to differentiate respondents based on their type of disability. The data, however, had so few responses to these questions we were unable to examine potential differences across types of disabilities (see Shields, Schriner and Schriner 1998b).

On the other hand, recent changes in the question wording and available responses in the 1994 Current Population Survey-Voter Supplement (U.S. Department of Commerce 1994) permit some preliminary investigations into the voter turnout of people with disabilities; the 1994 CPS now permits identification of disability as a reason for lack of participation in the labor force. Unfortunately, respondents who are active in the labor force are not asked if they have a disability, and we were therefore unable to examine potential differences between employed and unemployed individuals with disabilities. Consequently, generalizations of the findings from this study must be confined to people with disabilities who currently are not in the labor force.

It is important to note, however, that studies consistently indicate that about two-thirds of people with disabilities are unemployed (National Council on Disability 1996). In addition, we also controlled for the effects of unemployment within our models of political participation. Thus, while we must limit our generalizations to those individuals with disabilities who are unemployed, controlling for the effects of unemployment permits us to be certain that any unique relationships found within this sample of Americans with disabilities does not result solely from their lack of participation in the labor force (see Shields, Schriner and Schriner 1998a).

Overall, the available data sets are far from optimal and leave open the important questions regarding how our findings might be different if we had better methods of defining and identifying people living with disabilities. While such concerns are vitally important to our research agenda, we save a discussion of these, and related issues, to the final section of this review.

THEORETICAL PERSPECTIVES ON POLITICAL BEHAVIOR: COSTS AND BENEFITS/MOBILIZATION THEORY

Theoretical accounts of political participation generally acknowledge that the decision to participate in political life is often a costly choice (Downs 1957; Jackman 1993). For example, Rosenstone and Hansen (1993, p. 10) argue that individuals "participate in politics when they receive valuable benefits that are worth the costs of taking part"—otherwise, they abstain. Studies of participation have often focused on factors that either decrease the costs or increase the rewards of participation and, therefore, influence decisions to take part in political life. These factors fall into three broad categories: first, resources available to individuals; second, institutional factors associated with the American electoral system; and third, the political contexts in which individuals and groups must operate (Rosenstone and Hansen 1993; Verba, Schlozman and Brady 1995; Wolfinger and Rosenstone 1980).

Resources Available to Individuals

Cost-benefit theory assumes that taking part in politics places demands on peoples' resources, and that, therefore, individuals will be more likely to participate in politics if they have more resources. Research demonstrates that participation in electoral and community politics varies with demographic, psychological, and sociological factors. For example, the wealthy and well-educated are among the most able to afford the costs of political participation (Rosenstone and Hansen 1993; Verba et al. 1995; Wolfinger and Rosenstone 1980).

Institutional Factors

Institutional variables also alter the dividends of political participation. One factor that strongly influences participation is state-level registration requirements. Research consistently demonstrates that states with more lenient registration requirements have higher turnout rates, while states with more restrictive registration requirements have lower turnout rates (Nagler 1991; Patterson and Caldeira 1983; Wolfinger and Rosenstone, 1980).

Political Context

The costs of political involvement also depend on the political environment. For example, in states where gubernatorial or senatorial races are held concurrently with presidential elections, the political spectacle generally is more intense. In these years, news coverage is more widespread and candidates become more visible. Since elections for governor and senator tend to be highly publicized and intensely fought campaigns, information about the election is more available and "cheaper" than in their absence (Caldeira, Patterson and Markko 1985; Gilliam 1985; Rosenstone and Hansen 1993). The result of the greater intensity of these additional elections is that political information is easier to obtain; and, therefore the costs of participation are substantially reduced, increasing the likelihood of participation.

A complementary theoretical framework used to explain political participation emphasizes variables in the political process itself. A growing body of work has begun to focus on the role of the political elite in mobilizing and recruiting ordinary citizens to participate in elections. Central to this framework are the efforts of political parties, candidates and other political groups to reach out to people in order to influence their decisions concerning whether to participate in politics and in what manner. Research indicates, however, that often such mobilization efforts are biased toward the more affluent segments of society, and may even be strategically formulated to discourage "less desirable" groups from entering political life (Rosenstone and Hansen 1993; Verba, Schlozman and Brady 1995).

DATA AND METHODS

In the next section of this manuscript we review some of our own prior research regarding the political participation of people with disabilities. But first we briefly discuss the data sources and methods involved in these studies. The investigations summarized here utilized data from two primary sources. First, we conducted analyses of the 1987 Louis Harris and Associates survey of political participation of Americans with disabilities (see Shields, Schriner and Schriner 1998b). The survey asked participants about their involvement in the 1984 and 1986 elections. Second, we conducted analyses of the Current Population Survey-Voter Supplement national study conducted after the 1994 midterm elections (for a full report see Shields, Schriner and Schriner 1998a; for a description of variables see the Appendix).

Concerning the first study, the International Center for the Disability contracted with Louis Harris and Associates to conduct a survey of the political involvement of disabled individuals in America (see Louis Harris and Associates 1986). Between November 30 and December 23, 1985, 12,500 randomly chosen households were screened to identify individuals with disabilities. Telephone interviews with 1,000 non-institutionalized disabled persons age 16 and over were conducted to collect general demographic information. Following this initial survey, 536 of the original 1000 were reinterviewed in June and July of 1987 to record their political participation and involvement during the 1984 and 1986 elections.

While the Harris study allows us to explore many types of political attitudes and behaviors, it contains several weaknesses. Incomplete and missing data was a serious problem that prevented any thorough analyses across types of disabilities. Further, the sample size, while adequate, is not tremendously large—causing some concern about the precision of the parameter estimates. As a result of these problems, we have also utilized the 1994 Current Population Survey, November Voter Supplement.

The population study contains complete data from over 80,000 adults of which over 3,000 identify themselves as not working due to a disability. Further, this extremely large data base was obtained through a random sample of all Americans from every state. These strengths are the main reasons the CPS has been, and continues to be, a primary source for research for scholars of American political behavior. While the CPS does provide us with great confidence in the precision of the parameter estimates, as mentioned previously, the data set is far from ideal.

The greatest problem, for our purposes, is that people living with disabilities were identified through a question asking if and why they had been absent from the labor force. From this question we are able to identify over 3,000 people who are not in the labor force due to disability, but we are unable to identify any people living with disabilities who are active in the labor force. Consequently, while the CPS overcomes many of the problems associated with the Harris study, the findings must not be generalized beyond those people with disabilities who are not in the labor force.

In terms of methods, we rely largely on logistic regression. As many methodologists have demonstrated, ordinary least squares regression techniques are not appropriate when analyzing data with dichotomous dependent variables. Since the dependent variable we examine is dichotomous (did a participant vote or not) any reliance on ordinary least squares regression is inappropriate (King 1989). Very generally, logistic regression is a maximum likelihood technique that restricts predicted probabilities to lie between zero and one (in this case representing abstention and voting, respectively). Many functional forms meet the requirements for dichotomous dependent variable regression. Among the most popular, however, are the logit and probit functional forms. Since the logit functional form is mathematically more tractable, and because probit analysis and logistic regression produce nearly identical inferences and substantive conclusions, we opt for the more simple logistic functional form. Logistic regression is typically included in most standard statistical packages and is a widely accepted research methodology in studies of political behavior (King 1989).

RESULTS

Looking first at the descriptive data from the 1987 Harris study, as shown in Table 1, we see that nearly 66 percent of people with disabilities reported that they voted in the 1986 midterm elections and nearly 75 percent reported voting in the 1984 presidential election.

We are concerned that these estimates probably represent a good deal of "over-reporting." We believe that many people in this survey either incorrectly remembered their voting status (thought they had voted in these elections, but in fact they had not) or they fell prey to the pressures

Table 1. Percentages of Voting among People with Disabilities in the 1984, 1986, 1994 Elections

	1984 Harris Poll (1)	1986 Harris Poll (1)	1994 Current Population Study (2)
Voted	74.61	65.90	33.1 (53.7)
Didn't Vote	25.39	34.10	66.9 (46.3)
Sample size	512	525	8,2047

Notes: Cell entries are percentages. Percentages marked in parentheses are corresponding turnout figures for people who are not out of the labor force due to a disability.
See the section on "Defining Disability" in this chapter for a discussion of the many ways disability was identified in this study.
These participants are people who are out of the labor force as a result of a disability.
Source: Adapted from Shields, Schriner, and Schriner 1998a.

of social desirability and reported voting when they did not. In fact, overreporting of voter turnout continues to be one of the most difficult aspects of accurately measuring voter turnout in America. After all, few people are eager to admit that they have not voted in national elections.

Comparatively, as reported in Table 1, of the 80,000 nondisabled people sampled in the 1994 CPS, 42,261, or nearly 54 percent, reported that they voted, while only 1,111, or 33.1 percent of people with disabilities reported voting in the midterm elections—a difference of over 20 percentage points.

These lower turnout figures from the CPS, we feel, are more precise than the estimates from the Harris studies. The larger sample size and the experience of the Census Bureau suggest that these figures are likely to be a more accurate portrayal of the actual voter turnout of people with disabilities. Nevertheless, it is a very real possibility that there was a greater degree of voter turnout among people with disabilities in the 1980s compared to the 1994 election. While such questions are beyond the scope of available data, such possibilities remain. We must also remember that the CPS measure of disability is not perfect, and it could be that this lower estimate is a result of the question rather than changes in the participation patterns of people with disabilities.

Regardless of these questions, overreporting of voter turnout, as well as measurement problems in general, continued to be the bane of all scholars of electoral participation. Nevertheless, much research has concluded that overreporting is equally likely across many categories of socioeconomic groups. Consequently, multivariate analysis of the indi-

Table 2. Political Participation of People with Disabilities in 1984 and 1986 (Lou Harris Survey)

	Turnout in 1984?	Turnout in 1986?
Demographics and Resources		
Intercept	-3.521(2.67)*	-7.686(2.69)***
South	0.330(.451)	-0.289(.448)
Metropolitan area	-0.516(.474)	-0.372(.446)
Unemployed	-0.305(.501)	-0.459(.478)
Church member	0.267(.471)	0.070(.488)
Married	0.077(.522)	0.698(.512)*
Education	0.014(.122)	0.052(.115)
Income	0.082(.196)	0.157(.188)
African American	-0.794(.649)	-1.074(.718)*
Woman	-0.433(.427)	-0.425(.423)
Age	0.064(.072)	0.179(.073)***
Age^2	-0.000(.001)	-0.002(.001)***
Psychological Engagement		
Interest	1.130(.308)***	1.568(.321)***
Trust	-0.001(.321)	0.568(.319)**
External efficacy	-0.032(.441)	0.506(.437)
Internal efficacy	0.845(.576)*	-0.098(.533)
Disability Related Factors		
Group consciousness	-0.319(.221)*	-0.002(.213)
Retrospective evaluations	0.138(.149)	-0.022(.143)
Perceive people with disabilities as a minority group?	-0.786(.439)**	0.195(.422)
Age of onset of disability	0.088(.207)	-0.199(.199)
Severity of disability	-0.126(.217)	-0.208(.214)
N	201	204
-2 log likelihood	169.858	176.225
% correctly predicted	76.1	73.5

Notes: Cell entries are logistic regression estimates with standard errors in parentheses.

*$p<.10$, **$p<.05$, ***$p<.01$; one-tailed tests.

Source: Adapted from Shields, Schriner, and Schriner 1998b.

vidual-level determinants of voter turnout are often very accurate despite widespread overreporting of electoral activity (see Rosenstone and Hansen 1993, pp. 58-59). Therefore, with precautions in mind, we turn to a review of our findings concerning the factors predicting voter turnout among people with disabilities as measured by both the Harris and CPS surveys (see Shields, Schriner and Schriner 1998a, 1998b; Schriner, Shields, and Schriner 1998).

MULTIVARIATE RESULTS

Shown in Table 2 are some of our findings from the 1987 Harris study. Our most important discovery from this analysis was the consistent importance of being interested in politics. Those people who had disabilities and were not interested in politics were substantially less likely to have voted in either 1984 or 1986—even controlling for all the other factors included in the model. In fact, while we do not show the findings here, we found political interest to be a consistent and powerful positive factor for predicting many different forms of political activity among people with disabilities (see Shields, Schriner and Schriner 1998b). Perhaps most striking about these findings is that the effect of political interest was more consistent and, in most cases, more powerful than standard demographic variables such as income and education—both of which have been considered to be among the most important factors for predicting voter turnout among people without disabilities (Rosenstone and Hansen 1993).

Overall, the secondary analysis of the Harris survey suggests that political interest among people with disabilities is one of the most predictive factors involved in political participation. Of course, we must keep in mind that the ultimate utility of the survey is limited by the small sample size and gaps in the data set that prevented analysis of many respondents with missing data. If however, these findings are accurate, it would appear that political elites may have a strong impact on the extent of political involvement of people with disabilities, depending on the degree to which election campaigns interest these people, or to the extent that politics in general appeals to this minority group. On the other hand, to the degree that political elites ignore people with disabilities, we would expect their participation rates to drop. We return to these possibilities in the conclusion of this review.

Examining the individual-level determinants of voter turnout in 1994 (using the CPS), we present a model based on our previous work (see Shields, Schriner and Schriner 1998a). Focusing primarily on the disability dichotomous variable and the interactive effects in the model (which indicate the impact of the independent variable among people with disabilities), we first see that the main effect of being disabled is negative and significant—an unemployed person with a disability was significantly less likely to vote in the 1994 election—even when controlling for the effects of socioeconomic factors.

Table 3. Demographic and Institutional Model of Voter Turnout in the 1994 Congressional Elections, Current Population Survey

	Voter Turnout
Constant	-5.22 (.077) ***
South	-.281 (.018) ***
Education	.394 (.007) ***
Family Income	.059 (.003) ***
Age	.072 (.003) ***
Age^2	-.0004 (.00003) ***
African American	.313 (.028) ***
Married	.288 (.018) ***
Metropolitan Area	-.165 (.018) ***
Male	-.048 (.016) ***
Senate Race	.093 (.018) ***
Gubernatorial Race	.175 (.018) ***
Closing Date	-.006 (.0009) ***
Years in the Community	.288 (.006) ***
Unemployed	-.034 (.067)
Age * Unemployed	-.003 (.002) **
Person with a Disability	-1.29 (.455) ***
Person with a Disability * Family Income	-.02 (.012) *
Person with a Disability * Age	.061 (.017) ***
Person with a Disability * Age^2	-.0007 (.0001) ***
Number of Cases	80,409
Percent correctly predicted	70.1
-2 Log Likelihood	92554.282

Notes: Cell entries are logit estimates with standard errors in parentheses.
* $p<.10$, ** $p<.05$, *** $p<.01$; two-tailed test.
Source: Adapted from Shields, Schriner, and Schriner 1998a.

In addition, while the overall effect of income is positive, the effect is significantly attenuated among those who are not in the labor force because of a disability.

The effects of age are dramatically different among people with and without disabilities. These interactive relationships are graphically depicted in the conditional probability plot (see King 1989) presented in Figure 1. The scenario assumes the "typical" voter from the 1994 sample. As the graph depicts, the person who has a disability is less likely than the nondisabled person to vote at all age levels.

The person with a disability, however, is much less likely to vote than the nondisabled person as age increases. The effects of age appear to increase the probability that a nondisabled person will vote at all ages, albeit with a more attenuated effect once this person reaches the most

Disenfranchised

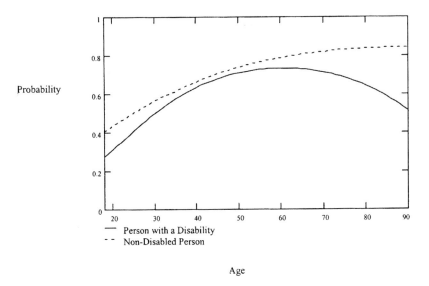

Figure 1. Effects of Age on Voter Turnout in 1994 Contrasting a Person with a Disability and a Non-Disabled Person

advanced ages. The effects of an additional year of age for the nondisabled person, even during the oldest years, are associated with a positive increment in the probability that this person will vote. These are similar to findings reported by Rosenstone and Hansen (1993), who found evidence supporting the life-experience hypotheses of aging—that political involvement increases across the entire life span.

Comparatively, however, the positive effects of age for the person with a disability begin to level off in midlife and then begin a rather steep decline. For example, a 25-year-old nondisabled person and a 25-year-old person with a disability have a probability of voting of 50 percent and 41 percent, respectively. These probabilities, however, change to 71 percent and 67 percent, respectively, at the age of 45 and then become 82 percent and 69 percent at the age of 70. Across the 25 years from ages 45 to 70, the probability of voting for a nondisabled person increases by 11 percentage points, while the probability of voting for a person with a disability increases only 2 percentage points. These findings indicate that age appears to play a unique and dramatic role in affecting the likelihood of voter turnout among people with disabilities.

We must note that cross-sectional surveys provide a rather blunt instrument when it comes to disentangling the effects of age. In fact, cross-sectional data do not permit us to disentangle the effects of three different age-related factors: the influence of cohorts, period effects and maturation.

Generally speaking, this socialization period refers to childhood and other "early" experiences. For example, great differences exist in the early socialization, and therefore political participation patterns, of "Generation X" versus people who were socialized during the Great Depression. Period effects refer to political events and occurrences, generally taking place later in life, that have a great influence on a particular segment of the population—such as the effects of the Vietnam War among young people during the 1960s—or the effects of the civil rights movement among young minorities. The effects of maturation refer to the psychological, biological and social changes that occur as a result of simply growing older. In cross-sectional surveys all these influences are confounded and nearly impossible to disentangle—and all measured by the single indicator of age.

Nevertheless, it is our interpretation that the impact of age among people with disabilities is that elderly people with disabilities are less likely to participate in political life than elderly people who do not have disabilities. Of course, until longitudinal data are gathered, these conclusions remain tentative.

The CPS data also allow us to examine variations in political participation patterns as a function of gender and race (see Schriner, Shields and Schriner 1998). As shown in Table 3, we have added to the model of voter turnout all possible interaction terms between the variables named "Woman," "African American" and "Person with a Disability" (which indicate the influence of the independent variable among women, and African Americans and people with disabilities).

Looking first at the three-way interaction term between Woman, African American, and Person with a Disability we see that this variable is negative but insignificant—indicating that African American women with disabilities were statistically no more or less likely to have voted compared to Caucasian non-disabled men, once we control for demographic factors, such as education and income.

The two-way interaction term between African American and Person with a Disability is also insignificant, indicating that African American males with a disability were insignificantly different from nondisabled

Table 4. Predicting Registration and Voter Turnout in the 1994 Congressional Elections: Comparing Race, Gender and Disability Current Population Survey

	Voter Turnout
Constant	-5.27 (.078) ***
South	-.281 (.018) ***
Education	.394 (.007) ***
Family Income	.059 (.003) ***
Age	.072 (.003) ***
Age^2	-.0004 (.00003) ***
African American	.224 (.045) ***
Married	.289 (.018) ***
Metro	-.164 (.019) ***
Woman	.037 (.017) **
Senate Race	.093 (.019) ***
Gubernatorial Race	.175 (.018) ***
Closing Date	-.006 (.001) ***
Years in the Community	.288 (.006) ***
Unemployed	-.035 (.067)
Unemployed*Age	-.003 (.002) **
Person with a Disability	-1.26 (.461) ***
Person with a Disability*Family Income	-.021 (.012) *
Person with a Disability * Age	.06 (.017) ***
Person with a Disability * Age^2	-.0007 (.0001) ***
Additional Interactions to Explore Race and Gender Differences	
Person with a Disability * African American	.087 (.15)
Person with a Disability*Woman	-.065 (.093)
Woman *African American	.149 (.057) ***
Person with a Disability *Woman*African American	-.114 (.198)
Number of Cases	80409
Percent Correctly Predicted	70.09
- 2 log likelihood	92546.644

Notes: Cell entries are logistic regression estimates with standard errors in parentheses.
  *p<.10, ** p<.05, *** p<.01
Source: Adapted from Schriner, Shields, Schriner 1998

Caucasian males in terms of their likelihood of voting in the 1994 elections once we control for socioeconomic factors. In other words, African American men and women who have disabilities had probabilities of voting that were statistically indistinguishable from nondisabled Caucasian males once we control for the fact that African Americans (those with and without disabilities) have fewer resources than their Caucasian counterparts.

The interaction term between Woman and African American, however, is positive and significant—indicating that once we control for the lower socioeconomic status of nondisabled African American women, they were actually more likely to have been registered and to have voted than were nondisabled Caucasian men. Put differently, a nondisabled African American woman who has a socioeconomic base similar to a nondisabled Caucasian male was more likely to have been registered and to have voted in the 1994 elections.

The main effect of the variable "Woman" is positive and significant, indicating that nondisabled Caucasian women were more likely to have registered and to have voted than nondisabled Caucasian men. Further, the main effect of the variable "Disabled" is negative and significant indicating that Caucasian men with disabilities were less likely to have participated than nondisabled Caucasian men.

Unfortunately, what our data cannot answer is why Caucasian men and women with disabilities were less likely to vote, once we control for their socioeconomic status. One possible explanation, however, is the importance of the church in the political mobilization of African Americans (Verba, Schlozman and Brady 1995).

DISCUSSION, IMPLICATIONS, AND DIRECTIONS FOR FUTURE RESEARCH

The studies reviewed here offer a first-glimpse picture of the electoral participation of America's largest minority group. Results from these studies suggest, first, that people with disabilities may be much less likely to take part in the country's political life than nondisabled individuals. Voting rates that are 20 percentage points lower than those of the general population (as measured by the CPS), strongly suggest that disability is a demographic variable that should be, at the very least, included in future analyses of political behavior—especially when we note that demographic characteristics such as education, income and employment have less influence on the political participation of people with disabilities than for nondisabled individuals.

These results also cast doubt on the conventional wisdom that all people are more likely to participate as they age. Our findings—that the probability of voting among persons with disabilities falls precipitously after middle age—raise questions about what role disability plays in shaping political participation across the age span. Why does participation begin to decline so dramatically in mid-life? Are these declines due to attenuated connections to community life resulting from absence from the workforce and other societal settings; the difficulties posed by decreasing capacity to move about and communicate with others or generational differences in attitudes toward political participation? Or does participation fall because people with disabilities internalize the messages conveyed to them by public policy—that their problems, concerns and needs are not as important as those of others? The data analyzed in these studies do not permit us to answer these questions, but we suspect that views of citizenship that may be adopted by people with disabilities as a byproduct of public policy play some role in shaping their political participation.

The findings from the analysis of race and gender differences are also suggestive of the need to take disability into account in theories of political participation. Our research indicates that both male and female African Americans with disabilities are more likely to participate in electoral politics than are white men and women with disabilities. These results, which raise the possibility that other factors associated with voter turnout among African Americans such as group consciousness or church membership (see Verba, Scholzman and Brady 1995), are important factors for the disabled members of these groups.

We must, however, be cautious when interpreting these findings. The studies we report here are simply the beginning of a long-term effort to understand the role of people with disabilities in the American electorate, and are far too incomplete to permit great confidence when making interpretations or conclusions. These studies do, however, raise important questions for the future study of political activity for people with disabilities.

Some of the more difficult issues include the most appropriate manner in which to measure disability. For example, we hope to address:

1. what unit of analysis to use (should the definition of disability be measured at the individual level, at the nexus between individual and environment, or solely at the environmental level?);
2. how to reflect the fluid and continuous nature of many "disabling" conditions (Zola 1993);
3. how to acknowledge the social constructions of disability in "objective" measures;

4. whether and how to permit the use of different definitions for different purposes (i.e., may we use varying definitions in legislation to identify target populations, but insist on developing a consensus on what "disability" is for research on, say, the incidence and prevalence of disabling conditions?).

At present, definitions *do* reflect different units of analysis, generally do *not* reflect the fluid and continuous nature of impairments, do *not* adequately acknowledge the social constructions underlying "objective" measures, and *are* tied to the purposes for which they are developed (i.e., the CPS measure is an attempt to identify the number of people out of the labor force as a result of disability).

In the modern American democracy, the statutory definitions of disability embed complex and diverse societal views toward individual differences. The group of individuals who are legally labeled as "disabled" includes persons with physical impairments such as spinal cord injury or muscular dystrophy, cognitive impairments such as mental retardation or autism, emotional impairments such as schizophrenia and sensory impairments such as blindness or deafness.

The 43 definitions of disability in federal law (Kemp 1991) are used to accomplish different policy objectives. The definition employed by the Social Security Administration, for example, is intended to identify those individuals who should be granted income replacement benefits because of their inability to participate in the workforce (Stone 1984). The Administration on Developmental Disabilities, however, uses a definition that is based on the acquisition of a disability before the age of 21 because of its history of providing services to persons with mental retardation. The Americans with Disabilities Act provides protections not only to individuals who have physical or mental impairments, but also those who have a "record" of having an impairment or who would be "regarded" by others as having an impairment, because they may experience discrimination.

As with statutory definitions, there is also variation in disability definitions used in other government surveys. The National Health Interview Survey estimates the number of individuals who have impairments "a loss of mental, anatomical, or physiological structure or function that may be caused by active disease, residual losses from formerly active disease, and congenital losses or injury not associated with active disease" (LaPlante 1991, p. 59) or activity limitations "in actions or activities that are due to such impairments" (LaPlante 1991, p. 59). The Survey of Income and Program Participation (SIPP) asks respondents to indicate if they have functional limitations "actions" such as seeing words or letters (LaPlante 1991, p. 60). The SIPP also asks whether respondents require assistance in performing activities of daily living, such as dressing and eating and instrumental activities of daily living, such as handling personal finances and taking medications (National Institute on Disability and Rehabilitation Research 1992). The Census Bureau employs a measure of work disability (National Institute on Disability and Rehabilitation Research 1992). The variability in definitions illustrates that disability is not easily defined and does not represent one or more obvious distinctions, but rather is constructed by social processes.

The debate about how to define disability—and for what purposes—is not likely to subside any time soon. The nature of this debate proves that disability—probably more so than other "master statuses" such as race and gender—is a highly contextual and value-laden term. The debate also illustrates important cleavages in our understanding of the origins of the disability category. For example, one influential theory holds that disability serves as a "keystone" in a welfare state in which a fundamental distinction must be made between those who have legitimate reasons for not participating in the workforce (the "deserving poor"), and those who do not (Stone 1984, p. 12). In Stone's formulation, the term "disability" emerged as a catchall label for those individual variations in physical and mental capacity (such as "lameness" and "feeble-mindedness") that "mattered" in a new and evolving capitalist economy. People with "limitations" in mobility or cognitive skills could be excused from the work role. Over time, "disability" has become a political category that signifies a range of meanings (and today, of course, many people with disabilities work). But by extension, this theory of the origins of the disability category implies that all of these meanings have roots in the use of "disability" as an administrative tool for identifying the "deserving poor."

Another prominent theory holds that disability is a sociopolitical category (like gender and race) that has been used to justify discriminatory practices and prejudicial attitudes (e.g., Gleidman and Roth 1980; Hahn 1985; Safilios-Rothschild 1970). From this vantage point, the life circumstances of people with disabilities are explained not by their impairments but by society's reactions to the impairments:

> Disabled citizens have confronted barriers in architecture, transportation and public accommodations which have excluded them from common social, economic and political activities even more effectively than the segregationist policies of racist governments ... Disabled individuals have been subjected not only to stereotyping, but also to stigmatizing, which has made them the targets of aversion and ostracism. Studies of public attitudes have revealed extensive intolerance of disabled persons which is related to indicators of discrimination against other minority groups (Hahn 1985, p. 93).

At this point, it is appropriate to repeat a conclusion made by Harlan Hahn and Paul Longmore (see Scotch 1988) that disability is "essentially determined by public policy. In other words, disability is whatever laws and implementing regulations say it is." The most important—perhaps the *only* really important—definitions for disability are those made through the political process.

This conclusion leads us back to the justifications for inquiring about the electoral participation of persons with disabilities. Through the electoral process, individual differences are identified (or not) as public problems, and are (or are not) targeted for stigmatization or normalization through public policy. Thus, political participation is critical for persons with disabilities to define themselves, their relation to their nondisabled fellow citizens, and their status as members of the polity.

Yet another emerging theoretical framework for explaining political participation is based on the social construction of groups that are targeted by public policy. This perspective places primary emphasis on the "messages" that public policy sends to subgroups within the polity. Because people with disabilities are, perhaps more than other minority groups, affected by governmental actions, the messages conveyed by these actions may be a factor in their decisions to participate or withdraw from political life. Social constructionists have argued that the way individuals and their concerns, issues and needs are represented in political discourse helps determine the nature and outcome of the debate. But social constructions of groups targeted by elected officials may also determine their perceptions of themselves as citizens, their perceptions of their problems and their patterns of political participation.

Schneider and Ingram define a social construction as "the cultural characterizations or popular images of the persons or groups whose behavior and well-being are affected by public policy" These characterizations, which emanate from media, literature, art, political discourse and public policy, are "normative and evaluative, portraying groups in positive or negative terms through symbolic language, metaphors, and stories" (Schneider and Ingram 1993, p. 334). Their theory, which build on earlier work explaining the process by which social problems are defined (Edelman 1988; Best 1989; Spector and Kitsuse 1987), emphasize that social constructions of groups are, to some extent, determined by public policy, and in turn, affect policymakers as they make public policy.

Schneider and Ingram argue that social constructions of target groups affect the *policy agenda* and the *design* of policy. Policies targeted to favored groups are more likely to be beneficial while policies directed to negatively constructed groups are more likely to be punitive. Policymakers are more likely to burden disadvantaged groups than favored groups. This treatment by policymakers is perpetuated by differences in the ability of target groups to influence electoral politics and the policymaking process. Advantaged groups have more abilities and resources to combat negative treatment by policymakers, while those with more negative social constructions have fewer resources and abilities to protect their interests.

The differences in political power interact with social constructions to produce different kinds of target populations. The model suggests that public policies directed to particular groups convey expectations about that group's problems and status as citizens. Policy sends messages about what government is supposed to do, which citizens are deserving (and which not) and what attitudes and participatory patterns are appropriate for the target population to adopt. Different target populations, however, receive quite different messages. Policies that have detrimental impacts on a target population, or are ineffective in solving important problems for the target populations may not produce citizen participation directed toward policy change because the messages received by these target populations encourage withdrawal or passivity. Other target populations, however, receive messages that encourage them to combat policies detrimental to them through various avenues of political participation. As Schneider and Ingram argue:

> Policy is an important variable that shapes citizen orientations and perpetuates certain views of citizenship that are in turn linked to differential participation among groups. Groups portrayed as dependents (e.g., the "deserving poor") or deviants (e.g., criminals) frequently fail to mobilize or to object to the distribution of benefits and burdens because they have been stigmatized and labeled by the policy process itself. They buy into the idea that their problems are not public problems, that the goals that would be most important for them are not the most important for the public interest, and that government and policy are not reme-

dies for them. They do not see themselves as legitimate or effective in the public arena, hence their passive styles of participation. In contrast, the advantaged groups (e.g., the elderly) are reinforced in pursuing their self-interests and in believing that what is good for them is good for the country. They can marshal their resources and use them to gain benefits for themselves, all the while portraying themselves as public-spirited. Others do not object, and in fact, support such policies, because they accept the goals that benefit the advantaged groups and believe these groups are deserving of what they get. Social constructions enhance their power, whereas it detracts from the power of the disadvantaged groups (Schneider and Ingram 1993, p. 344).

Schneider and Ingram thus place a great deal of attention on how individuals view "citizenship." Views of citizenship are the attitudes held by individuals toward themselves as targets of public policy and as citizens. This concept includes attitudes about whether one's problems are political or individual; attitudes about what constitutes responsible citizenship; attitudes about one's right to ask for government assistance; perceptions of societal benefits from helping one's group; and perceptions about the appropriateness, effectiveness and value of political participation.

These views of citizenship are derived from public policies directed at one's group, and have profound implications for the beliefs one has about the importance of one's problems, the sources of their solutions and the appropriate levels and types of participation strategies in the policy arena. Views of citizenship also derive from the perspectives of other citizens. The benefits and burdens of policy are accepted or rejected by other citizens depending on the positive or negative social construction of the target group for the policy. Most citizens are more likely to support more benefits and fewer burdens for advantaged groups (because of their greater political power and positive social construction) and fewer benefits and larger burdens for other groups (because they have less political power and a negative social construction). Although not yet subjected to empirical tests, we believe that these theoretical developments regarding how social constructions and political power are tied to public policy messages appear promising for understanding the political involvement of people with disabilities.

Although these questions are entirely beyond the scope of existing data sources, the answers not only will provide insights into the individual dynamics of political activity among the nation's largest minority group, but also promise to provide insights into the health of modern American democracy. The extent to which people with disabilities are represented in our governmental will rest, to a great extent, on the manner in which people with disabilities participate in our electoral process, the manner in which they are represented by themselves and by political elites and the manner in which public policy effectively represents their needs and desires. We hope to see these questions become a central concern of social scientists and anyone concerned with improving American democracy.

APPENDIX

Variable Description From Lou Harris Survey

South	Coded 1 for respondents living in the south and 0 otherwise.
Metropolitan area	Coded 1 for respondents living in metropolitan areas and 0 otherwise.
Age of onset of disability	Age in which disability began: 1= at birth, 2=0-11 years old, 3=12-19 years old, 4=20-39 years old, 5=40-55 years old, 6=over 55 years old.
Severity of disability	Respondent=s description of disability: 1=slight, 2=moderate, 3=somewhat severe, 4=very severe.
Unemployed	Coded 1 for respondents who were unemployed and zero otherwise
Group consciousness	Does respondent feel they have a sense of common identity with other people with disabilities: 1=no sense of common identity, 2= some sense, 3=somewhat strong sense, 4=very strong sense of common identity.
Retrospective evaluations	Does respondent think things have changed in general for Americans with disabilities in the past 10 years: 1=much worse, 2=somewhat worse, 3=no change, 4=somewhat better, 5=much better.
Minority group	Does respondent feel that people with disabilities are a minority group in the same sense as are black and Hispanics: 1=yes, 0=no.
Church member	Coded 1 for respondents who reported that they were members of a church or synagogue and 0 otherwise.
Married	Coded 1 for respondents who were married and 0 otherwise.
Education	1=no formal schooling, 2=first through seventh grade, 3= up to eighth grade, 4= some high school, 5=high school graduate, 6=some college, 7=two year college, 8=four year college, 9= postgraduate.
Income	1=$7,500 or less, 2=$7,501 to $15,000, 3=$15,001 to $25,000, 4=$25,001 to $35,000, 5=$35,001 to $50,000, 6=$50,001 or over.
African American	Coded 1 for respondents who identified themselves as African American and 0 otherwise.
Woman	Coded 1 for respondents who identified themselves as women and 0 otherwise.

(continued)

Appendix (Continued)

Interest	Respondent's reported interest in the most recent political campaign: 1=not much interest, 2= somewhat interested, 3= very much interest.
Trust	Respondent's trust in Washington government: 1=none of the time, 2=some of the time, 3=most of the time, 4=always.
External efficacy	People like me have no say in government. Coded 1 for respondents who disagree and 0 otherwise.
Internal efficacy	People like me can't understand government. Coded 1 for respondents who disagree and 0 otherwise.
Age	Respondent's age in years.
South	Coded 1 for respondents in southern states; 0 otherwise.
Education	Six categories ranging from 1= respondent completed less than 9th grade, to 6= respondent graduated from college.
Family Income	Fourteen categories ranging from 1= less than $5,000 per year, to 14= $75,000 or more.
African American	Coded 1 for African American respondents; 0 otherwise.
Married	Coded 1 for married respondents; 0 otherwise.
Metropolitan Area	Coded 1 for respondents in metropolitan areas; 0 otherwise.
Male	Coded 1 for male respondents; 0 otherwise.
Senate Race	Coded 1 for respondents from states with a senate race in 1994; 0 otherwise.
Gubernatorial Race	Coded 1 for respondents from states with a gubernatorial race in 1994; 0 otherwise.
Closing Date	Number of days before election that respondents were required to register to vote in 1994.
Years in the Community	Sixteen categories ranging from 1=less than 1 month, to 6=5 years or longer.
Unemployed	Coded 1 for respondents who were not in the labor force because (a) they were unemployed, (b) there were unemployed but looking for work, or (c) they were disabled; 0 otherwise.
Person with a Disability	Coded 1 for respondents who could not participate within the labor force because of a disability; 0 otherwise.
Voted	Coded 1 for respondents who reported voting in the 1994 election and 0 otherwise.
Registered	Coded 1 for respondents who reported being registered in the 1994 election and 0 otherwise.
Absentee Voting	Among those respondents who reported voting in the 1994 election, those who reported voting absentee were coded 1 and those who voted but did not vote absentee were coded 0.

Source: Adapted from Shields, Schriner, and Schriner 1998a.

NOTES

1. The nomenclature of disability is in transition, and reflects the difficulty of discussing human variation without using value-laden terminology. While we use the terms "impairments" and "disability" throughout this manuscript, we recognize that both terms are problematical. "Impairments" refers to individual differences that have become public problems (which are often also labeled "disabilities"); and implicitly recognizes those individual differences (such as left-handedness or requiring more sleep than "normal" to function, that have not been—or perhaps, have not yet been—labeled as public problems). "Disability," a term with negative connotations among the general public, is embraced in the disability rights community as a term of personal identity (as in "disability pride").

2. Barnartt defines disability as a master status resembling other master statuses such as race and gender. All these statuses, she notes, are incorrectly viewed as being clear, distinct and unchanging. However, there are important differences in the "timing of onset of the status, the permanency of the condition, socialization into the status, the lack of homogeneity of the physical conditions which produce that status and societal reactions to that status" (Barnartt in press, p. 4). Disabilities may be congenital or acquired. Some impairments are cured. Except in rare circumstances, people with disabilities are not socialized as "disabled" at an early age. There is significant heterogeneity in the disability community. Finally, there is a range of societal reactions to disability.

REFERENCES

Barnartt, S.N., and R. Scotch. (in Press). *Contentious Political Actions in Disability Community: 1970–1999.* Washington, DC: Gallaudet University.
Best, J. (ed.). 1989. *Images of Issues: Typifying Contemporary Social Problems.* New York: Aldine de Gruyter.
Braddock, D., and R. Hemp. 1996. "Medicaid Spending Reductions and Developmental Disabilities." *Journal of Disability Policy Studies* 7(1): 1-32.
Caldeira, G.A., S.C. Patterson, and G.A. Marko. 1985. "The Mobilization of Voters in Congressional Elections." *The Journal of Politics* 47: 490-509.
Downs, A. 1957. *An Economic Theory of Democracy.* New York: Harper and Row.
Edelman, M. 1988. *Constructing the Political Spectacle.* Chicago: University of Chicago Press.
General Accounting Office. 1996. *SSA disability: Program Redesign Necessary to Encourage Return to Work.* Washington, DC: General Accounting Office.
Gilliam, F.D. 1985. "Influences on Voter Turnout for U.S. House Elections in Non-Presidential Years." *Legislative Studies Quarterly.* 10: 339-351.
Gleidman, J., and W. Roth. 1980. *The Unexpected Minority: Handicapped Children in America.* New York: Harcourt Brace Jovanovich.
Hahn, H. 1985. "Toward a Politics of Disability: Definitions, Disciplines, and Policies." *The Social Science Journal* 22(4): 87-105.

Jackman, R. 1993. "Rationality and Political Participation." *American Journal of Political Science* 37: 279-290.

Kemp, E.J., Jr. 1991. "Disability in Our Society." Pp. 56-58 in *Disability and Work: Incentives, Rights, and Opportunities*, edited by C.L. Weaver. Washington, DC: The AEI Press.

King, G. 1989. *Unifying Political Methodology: The Likelihood Theory of Statistical Inference*. Cambridge: Cambridge University Press.

LaPlante, M.P. 1991. "The Demographics of Disability." In *The Americans With Disabilities Act*, edited by Jane West. New York: Milbank Memorial Fund.

Louis Harris and Associates. 1986. *The ICD Survey of Disabled Americans: Bringing Disabled Americans into the Mainstream*. New York: International Center for the Disabled.

———. 1987. *Participation in Voting and Elections by Disabled Americans*. New York: Author.

Mayo, H.B. 1960. *An Introduction to Democratic Theory*. New York: Oxford University Press.

Nagler, J. 1991. "The Effect of Registration Laws and Education on U.S. Voter Turnout." *American Political Science Review* 85: 1393-1405.

National Institute on Disability and Rehabilitation Research. 1992. *Digest of Data on Persons with Disabilities: 1992*. Washington, DC: National Institute on Disability and Rehabilitation Research.

National Council on Disability. 1996. *Achieving Independence: The Challenge for the 21st Century*. Washington, DC: National Council on Disability.

Nosek, M.A., M.J. Fuhrer, D.H. Rintala, and K.A. Hart. 1993. "The Use of Personal Assistance Services by Persons With Spinal Cord Injury: Policy Issues Surrounding Reliance on Family and Paid Providers." *Journal of Disability Policy Studies* 4(1): 89- 103.

Patterson, S.C., and G.A. Caldeira. 1983. "Getting Out the Vote: Participation in Gubernatorial Elections." *American Political Science Review* 77: 675-689.

Percy, S.L. 1989. *Disability, Civil Rights, and Public Policy: The Politics of Implementation*. Tuscaloosa, AL: University of Alabama Press.

Pitkin, H.F. 1967. *The Concept of Representation*. Berkeley: University of California Press.

Rosenstone, S. J., and J.M. Hansen. 1993. *Mobilization, Participation, and Democracy in America*. New York: Macmillan.

Safilios-Rothschild, C. 1970. *The Sociology and Social Psychology of Disability and Rehabilitation*. New York: Random House.

Schneider, A., and H. Ingram. 1993. "The Social Construction of Target Populations: Implications for Politics and Policy. *American Political Science Review* 87: 334-347.

Schriner, K.F., and A.I. Batavia. 1995. "Disability Law and Social Policy." In *Encyclopedia of Disability and Rehabilitation*, edited by A.E. Dell Orto and R.P. Marinelli. New York: Macmillan.

Schriner, K.F., P. Rumrill, and R. Parlin. 1995. "Rethinking Disability Policy: Equity in the ADA Era and the Meaning of Specialized Services for People with Disabilities." *Journal of Health and Human Resources Administration* 17: 478-500.

Schriner, K.F., T.G. Shields, and K. Schriner. 1998. "The Effect of Gender and Race on the Political Participation of People with Disabilities in the 1994 Mid-term Election." *Journal of Disability Policy Studies* 9: 53-76.

Scotch, R.K. 1984. *From Good Will to Civil Rights*. Philadelphia, PA: Temple University Press.

_____. 1988. "Disability as the Basis for a Social Movement: Advocacy and the Politics of Definition. *Journal of Social Issues* 44(1): 159-172.

Shields, T., K.F. Schriner, and K. Schriner. 1998a. "The Disability Voice in American Politics: Political Participation of People with Disabilities in the 1994 Election." *Journal of Disability Policy Studies* 9: 33-52.

_____. 1998b. "Influences on the Political Participation of People with Disabilities: The Role of Individual and Elite Factors in 1984 and 1986." *Journal of Disability Policy Studies* 9: 77-89.

Spector, M., and J.I. Kitsuse. 1987. *Constructing Social Problems*. New York: Aldine de Gruyter.

Stone, D.A. 1984. *The Disabled State*. Philadelphia, PA: Temple University Press.

U.S. Department of Commerce, Bureau of the Census. 1994. *Current Population Survey, November 1994: Voting and Registration and Computer Usage* [Computer file].Washington, DC: U.S. Department of Commerce, Bureau of the Census [Producer]. Ann Arbor, MI: Inter-university Consortium for Political and Social Research [Distributor].

Verba, S. 1996. "The Citizen Respondent: Sample Surveys and American Democracy." *American Political Science Review* 90: 1-8.

Verba, S., K.L. Schlozman, and H.E. Brady. 1995. *Voice and Equality: Civic Voluntarism in American Politics*. Cambridge, MA: Harvard University Press.

Watson, S.D. 1993. "A Study in Legislative Strategy: The Passage of the ADA." In *Implementing the Americans With Disabilities Act*, edited by L.O. Gostin and H.A. Beyer. Baltimore, MD: Paul H. Brookes.

Wolfinger, R.E., and S.J. Rosenstone. 1980. *Who Votes?* New Haven, CT: Yale University Press.

Zola, I.K. 1993. "Disability Statistics, What We Count and What It Tells Us: A Personal and Political Analysis." *Journal of Disability Policy Studies* 4(2): 9-39.

STADIUM SIGHTLINES AND WHEELCHAIR PATRONS
CASE STUDIES IN IMPLEMENTATION OF THE ADA

Sanjoy Mazumdar and Gilbert Geis

ABSTRACT

Title III of the 1990 Americans with Disabilities Act was passed to end discrimination against persons with disabilities, to provide "functional equality" and to make buildings accessible. This public policy was aimed at integrating persons with disabilities into the mainstream. But, persons with disabilities, "social watchdogs" and governmental agencies have had to go to court to ensure compliance from private and governmental organizations charged by law to supply the services.

Through a study of two court cases, one involving the MCI Center in Washington, D.C., and the second the Rose Garden in Portland, Ore., we focus on public policy and its effectiveness and reach two major conclusions. First, we suggest that American public policy using law as an instrument can lead to vagueness in its formulation and ambiguity in its implementation. Second, we highlight the lessons that can be learned from a review of these court decisions and argue that persons concerned with shaping public policy have to attend assiduously to clarity in formulation of the law, the manner in which courts interpret laws and administrative guidelines, since this is often as much a political process as rote application of juridical principles and precedents.

INTRODUCTION

In the United States, the tripartite division in the formulation and implementation of policy—through elected members of Congress formulating the law, more permanent and appointed government agencies adding detailed regulations and courts ruling on disputes—leaves much uncertainty.

Members of the U.S. Congress often enact regulatory legislation that includes a declaration of noble intent and some broad guidelines that are vague and sometimes ambiguous (Edelman 1964). The reasons for the failure to attend to specific details are understandable. Lawmakers have neither the time nor the expertise to deal with each aspect of what sometimes are extraordinarily comprehensive and complex matters. Also, if the thrust of the law is made too precise there is an increasing chance that a detail will arouse opposition among lawmakers who are willing to take a general stand but might be opposed to specific elements if spelled out in the proposed statute. In addition, they have neither sufficient information nor prescience to anticipate problems that will arise when their legislated mandate has to be translated into operative public policy.

Legislators presume that government agencies entrusted with the task, and those charged with enforcing the law will provide regulatory standards that address with some exactitude what must be done to comply with the general requirements. In the event that these agencies fail to rise satisfactorily to the challenge of specificity, legislators assume that the courts, which cannot avoid an issue by inaction or indecision, will rule on arguable matters (Manning 1996). If the legislators do not approve the path the courts take, they can enact a new statute that remedies what they see as the unacceptable judicial opinion.

Judges, of course, can employ various obfuscatory tactics or other stratagems that can keep matters uncertain, but in doing so they invite criticism from the cadre of academics who make it a business to second guess the judiciary. They also run the risk of encouraging additional litigation that merely creates more business for them and ultimately forces them or their successors to obtain closure on the controverted issue.

For this project we focus on public policy with several research questions in mind. Public policy proponents advocate use of laws and other instruments to correct social problems and ills. But how well does the Americans with Disabilities Act (ADA) work as a public policy? In terms of effectiveness in achieving its stated aims, how does the ADA function? We examine the movement to translate the sometimes amorphous ingredients of ADA into specific public policy. How well does the description of the process detailed above fit the formulation, operation and implementation of the ADA? How do common folk deal with public policy designed by the highest lawmaking body of the land? We consider what lessons and implications our study has in regard to recourse to the courts for interpretation of the ADA statute and the derivative guidelines that specify with more precision what the law requires.

We will employ a case study methodology. A methodological virtue of the case study approach is that it allows the testing of speculative ideas against actual events; it offers data that will support or disconfirm intuitive conclusions (Eckstein 1975; Feagin, Orum, and Sjoberg 1991; Ragin and Becker 1992). Case studies cannot, of course, definitively prove a social scientific thesis, but they can provide invaluable insight for theory building and for the sharper reformulation of existing ideas. In addition, a case study enables examination of processes and details of actions and events. Case studies can provide valuable depth of detail. By concentrating on elements of a statute and the relevant implementing guidelines we obtain an appreciation of the raw material available to the judges to fashion their opinions.

More specifically, our case studies indicate the richness of material that can be found in court cases through a review not only of the decisions themselves, but of the briefs by opposing parties, their public statements, and commentaries by scholars and mass media.

Through studies of two major court cases, the MCI Center in Washington, D.C., and the Rose Garden in Portland, Ore., we shall provide an understanding of the implementation of public policy. As we point out in more detail in our conclusion, statutes often represent only a skeletal framework for policy, and the manner that policy comes to be implemented can be strongly influenced by how judges decide to interpret legislative language, which may be vague and accorded any one of several specific meanings. These litigations therefore highlight the necessity for persons with disabilities and those supporting their crusade for equality to attend very carefully and forcefully not only to lawmaking but to the processes that ensue in the wake of the passage of statutes.

The ADA, enacted in July 1990 to be effective from January 26, 1993, offers a quintessential illustration of the process outlined above. The Act, as two commentators have noted, is "characterized by high policy ambiguity and low policy conflict" (Condrey and Brudney 1998, p. 39), the latter referring to the absence of any significant amount of dissent from the law's general aims, as detailed below.

The purpose of Title III of the ADA dealing with accessibility to premises, the segment with which this paper is concerned, is boldly set forth in § 302(a):

> No individual shall be discriminated against on the basis of disability in the full and equal enjoyment of the goods, services, facilities, privileges, advantages, or accommodations of any place of public accommodation by any person who owns, leases (or leases to), or operates a place of public accommodation (42 U.S.C. § 12182 (a)).

In earlier days, persons with disabilities were condescended to and shut out of participation in activities routinely enjoyed by those without disabilities (Altman 1981; Fine and Asch 1988). In the words of one commentator, persons with disabilities ultimately "came out of the back bedroom" (Hahn 1993, p. 745), joining blacks, women, gays and other disesteemed groups in the battle to gain recognition and rights.

CASE I: THE MCI SPORTS ARENA, WASHINGTON, D.C.

The court fight over the architectural planning of the MCI sports arena to be constructed in downtown Washington, D.C., serves as one of the two case studies we will employ to provide details about how the ADA has served as a weapon by means of which persons with disabilities demanded the rights they maintained were granted to them by legislative mandate.

The MCI Center, located in downtown Washington, is the home of the Washington Wizards of the National Basketball Association, the

Washington Capitals of the National Hockey League, and will also serve as a multipurpose facility hosting other sports and entertainment events.

U.S. District Court, Washington, D.C. (June 1996)

The Paralyzed Veterans of America (PVA), located in Washington, filed suit in June 1996 in the federal district court in the District of Columbia (*PVA* v. *EBAE* 1996a) to enjoin the implementation of the architectural design of the MCI Center on the ground that it did not adequately meet the requirements of the Title III of ADA (42 U.S.C. § 12181 et seq.). The suit was part of the PVA "social watchdog" effort "to eliminate discrimination against its members on the basis of their physical disabilities" (*PVA* v. *EBAE* 1996a: 4t Plaintiff's Complaint). It should be noted that private parties, such as the PVA, cannot seek compensation for damages due to violations of Title III; they can only look for injunctive relief as a route to remedial action. The PVA is a nationwide organization chartered by the Congress just after World War II (36 U.S.C. § 1151 et seq.). Its approximately 17,000 members are almost exclusively veterans who served in the armed forces and have either spinal-cord injuries or spinal-cord disease. The group has 34 chapters across the nation; there are 345 members in the metropolitan Washington, D.C., branch.

The Seesaw Fate of Ellerbe Becket

The stadium had been designed by Ellerbe Becket Architects & Engineers, one of the largest architectural firms in the United States. Ellerbe Becket is incorporated in Delaware with Minneapolis as its principal place of business. While the focus of the suit was on the MCI Center, the plaintiff also pointed out that similar designs had been prepared by Ellerbe Becket or its subsidiaries for the FleetCenter in Boston, the Marine Midland Arena in Buffalo, N.Y., the Core States Arena in Philadelphia, and the Rose Garden in Portland, Ore. Lawsuits were filed in Buffalo, where an agreement between the parties ended the lawsuit, in Boston, by the local PVA, and in Philadelphia (see Conrad 1997). Litigation regarding the Portland arena forms the basis for our second case study.

At an early stage of the District of Columbia court proceedings, however, Ellerbe Becket was dismissed from the case on the ground that the phrase in the regulations—"a failure to design and construct facilities that are readily accessible"—must be read conjunctively, that is, a violator must have both designed and constructed the site in dispute. Other courts later would disagree with this interpretation. In January 1997, Jose A. Gonzalez, Jr., a U.S. district judge in Florida, adopted a position opposite to that of the D.C. court in a suit concerning plans for the Broward Arena, the future home of the Florida Panthers hockey team. Gonzalez preferred the plaintiffs' argument that if architects were not liable under the ADA, it was conceivable that no entity whatsoever would be liable for construction of a new commercial facility that violated the ADA (*Johanson v. Huizenga Holdings* 1997; see also *Caruso v. Blockbuster Sony Entertainment* 1999).

Nor did Ellerbe Becket's legal woes end with these cases. In October 1996, the U.S. Attorney General's Office in Minneapolis filed a complaint against the company on the ground that it "repeatedly designed arenas and stadiums with wheelchair seating locations that do not provide wheelchair viewers with lines of sight to the floor or field that are comparable to those of other spectators" (*U.S. v. EBI* 1997a: 1264). Two years later, a consent decree ended that case, with Ellerbe Becket agreeing that it would include adequate wheelchair seating and satisfactory lines of sight for all fixed seating facilities that it designed subsequent to the 1998 date of the consent order and would provide annual reports of compliance for three years. In exchange, the government agreed not to take any action against Ellerbe Becket for projects designed prior to 1998 (*U.S. v. EBI* 1998).

"We are not trying to get them to redesign the stadiums," a Justice Department spokesperson noted, "we're trying to ensure they no longer build stadiums that are inaccessible to the handicapped" (Bowles 1996). For a PVA attorney, the agreement with Ellerbe Becket marked "a sea change in focus and attitude of the firm," though he cautioned that "whether that permeates [the architectural world] is an open question" (Conrad 1998, p. 8).

The dismissal of Ellerbe Becket in the MCI case left as defendants the D.C. Arena Limited Partnership, a corporation formed to oversee the construction of the MCI Center and the Centre Group Limited Partnership, which would be responsible for operating the facility, as well as

Abe Pollin Sports, Inc. and Abe Pollin, the last two parties involved in the ownership and operation of the facility.

Lines of Sight and Standing Spectators

In its suit, the PVA asserted that at crucial moments in sporting events, able-bodied spectators would stand in order to more closely follow the action and thereby block the view of patrons in wheelchairs. This lack of provision of unobstructable lines of sight (ULOS), it claimed, violated a requirement set out in the Standards for New Construction and Alterations (also known as the ADA Accessibility Guidelines (ADAAG): see 28 C.F.R. § 36.406) promulgated by the federal Department of Justice (DOJ) to explicate provisions of Title III of the ADA, which is largely derived from the Architectural Barriers Act of 1968 (42 U.S.C. §§ 4151-4157). Section 4.33.3 of the guidelines, enacted in July 1991, declares that "[w]heelchair areas ... shall ... provide people with physical disabilities a choice of admission prices and lines of sight comparable to those for members of the general public ..."

Another section of the guidelines indicates the number of accessible wheelchair seats required, a formula of "1 percent plus one" of the total seating capacity (*PVA* v. *EBAE* 1996k: 398). The PVA claimed that all accessible seats are required to have unobstructed sight lines. The PVA pointed out that persons in wheelchairs could obtain unobstructed-view seats only in a few specific locations in the front row on the floor of the arena, or at the east and west end of the last row of the main concourse.

The failure of the designers and builders to provide wheelchair seating with adequate lines of sight dispersed throughout the arena, they claimed, deprived wheelchair patrons of a choice of admission prices comparable to those available to others, a matter said to be a violation of ADA.

The regulations also required that at least one companion fixed seat be available next to each wheelchair place. At this arena, all specified wheelchair seating was located directly behind fixed places for ambulatory spectators.

The MCI design also contained a provision for what was labeled "in-fill seating." This involved seating locations that could be configured either for no more than four persons in wheelchairs or as many as 27 ambulatory spectators. The in-fill section was not flexible, so that if one wheelchair occupant was seated in it, it could not be used to accommodate ambulatory spectators. There were 24 "in-fill" locations in the plan. The PVA asserted that economic pressures would lead the arena owners to steer wheelchair patrons into those "in-fill" areas already accommodating wheelchairs, so that they would "fail to provide accessible wheelchair seating in a setting that is an integral part of the MCI Center's fixed seating plan, in violation of ADA ..." (*PVA v. EBAE* 1996a: 21t[1] Plaintiff's Complaint).

In addition, the plan called for 109 suites with "living room areas" on two levels, having a total seating capacity of approximately 1600 (*PVA v. EBAE* 1996k: 397 TFH MO 20 Dec. 1996: 8-9t). The suites generally could accommodate up to 24 persons and would be leased as entities, usually to corporations, for terms from three to ten years at a cost of $100,000 to $175,000 a year each.

U.S. District Court, Washington, D.C., October - December 1996

In October 1996, Judge Thomas F. Hogan denied the defendants' motion for a summary judgment that would have dismissed the PVA case. He ruled that the law required "enhanced" (or ULOS) lines of sight for those in wheelchairs so that they can view the performance floor over persons in front who may be standing (*PVA v. EBAE* 1996i: 392 TFH MO 20 Dec 1996: 6t). He also ruled that wheelchair-accessible spaces had to be dispersed throughout the spectator seating bowl.

Two months later Judge Hogan issued a Memorandum Opinion. He began with a quotation from Alexis de Tocqueville (1868, p. 196) that focuses on the point raised in our introduction: "[T]here is hardly a political question in the United States that does not sooner or later turn into a judicial one." Hogan lamented that he had been forced to fill this void (*PVA v. EBAE* 1996k: 394).

The question of whether unobstructed sight lines over spectators standing in front was required was taken up by Judge Hogan. The ADA stated "full and equal enjoyment." DOJ, solely on the basis of an in-house decision, had in December 1994 added the present mandate to its regulations in a supplement to its *Americans with Disabilities Act Title III Technical Assistance Manual* (December 1994).

Judge Hogan bemoaned the lack of past or present guidance by the DOJ regarding the matters being litigated. He said that the department "has not seen fit to step up to its statutorily mandated role by providing concrete guidance for architects and builders" to help the court enforce the "demanding, and controversial design requirements that the Department of Justice has never championed in any court or in any rule making procedure ... and has declined to support in the present case, despite several invitations from the Court to do so" (*PVA v. EBAE* 1996k: 394).

Although the command of the rule was categorical, it had several elements that made it vulnerable to attack. Most importantly, it had not been, as is the procedure required by the Administrative Procedures Act, drafted, circulated for comment, and then, after a review of responses, adopted, reworded or dropped. Indeed, an earlier public inquiry by federal regulators about unobstructed viewing from wheelchairs had been challenged by many persons and the agency had declared that it would postpone any definitive resolution of the matter until a time when it issued regulations about recreational areas (56 Fed. Reg. 35,440 [1991]).

Judge Hogan castigated the DOJ, the agency entrusted the task of providing detailed guidelines. In short, "the ambiguity of the ADA regulations, and the lack of guidance and participation by the Justice Department in these matters," Hogan said, "has created an unfortunate situation in which defendants can act in good faith and still fail to comply with the law. It is," he declared, "a sad predicament." Later, the judge again would scold the DOJ for "foolishly" falling into a situation described centuries earlier by Thomas Hobbes (1651: pt. II, ch. 26): "The written laws, if they be short, are easily misinterpreted." Not content to let the matter rest there, he subsequently berated the DOJ for its failure to provide a "concrete workable record from which to discern a standard" but instead offering a "nebulous record, comprised mostly of informal documents, press releases, announcements, and correspondence" (*PVA v. EBAE* 1996k: 394). The result, he said, was that there existed "nothing brighter than speculative light" and a "hazy tapestry" of Justice Department "action and inaction" (*PVA v. EBAE* 1996k: 399). Even so, Judge Hogan then ruled in favor of the PVA (*PVA v. EBAE* 1996b).

Five disputed issues were pinpointed by the court:

1. the number of seats that must be accessible to persons with disabilities in wheelchairs;
2. the number that must provide enhanced sight lines;
3. whether operational measures could be introduced to create satisfactory sight lines;

4. the required dispersal of those seats;
5. whether spaces in the luxury suites could be counted toward the required number of enhanced sight line seats.

The court pointed out that at least the guidelines on the number of seats necessary for wheelchair accessibility offered a mechanical and clear formula: the "1 percent plus one" standard. Hogan declared that the MCI arena, designed to accommodate several events, had a maximum capacity of 17,989 for basketball layout, and according to that formula, 181 accessible locations were required for basketball games and a more or less similar total for other events, and that the MCI Center met this requirement by providing 189, 182, 213 and 165 accessible seats respectively for basketball, hockey, 360° end stage and 270° end stage layouts (*PVA* v. *MCI* 1996k:397 TFH MO 20 Dec 1996:8t).

On the second issue, Judge Hogan required the defendants to meet his level of satisfactory "substantial compliance." To achieve this in the Ellerbe Becket design of the MCI Center with only 71 locations with enhanced lines of sight available to wheelchair patrons, he required the addition of a specified number of unobstructed sight line seats, which would bring the number of such seats to 62 percent of required accessible locations for 270° end stage and 82 percent for hockey. Later, he approved a plan providing unobstructed view seats for 78 percent to 88 percent of the required wheelchair locations (*PVA* v. *EBAE* 1996k:403-404 TFH MO 20 Dec 1996:23t). Thus, on occasion, a judge granted the defendants the possibility of deviating from the requirements, apparently because it seemed to him reasonable, given competing demands on the architecture. Moved perhaps by the illogic of a ruling that would permit as many as 25 percent of wheelchair spectators to not be able to view the event when ambulatory spectators in front stood up, this segment of the court decision became the subject of a law review commentary and critique, advocating a policy of "rebuttable presumption" of 100% ULOS, where those who could not meet this requirement would have to demonstrate compelling structural or other reasons for that lack (Fritts 1998, p. 2654).

The defendants sought to persuade the judge that enhanced (ULOS) sight lines provided wheelchair users not with an equivalent, but rather a superior view than ambulatory spectators who lost some of their visibility when other spectators stood. The judge granted this claim, but observed that if those in front stand, an ambulatory spectator can stand

and regain most of his or her view. Similarly, as a later ruling would point out, children and short people can stand on their seats. "The court," Hogan noted, "is thus faced with a choice between requiring the superior, enhanced views or accepting the completely eclipsing, unenhanced views." It found that choice "an easy one," sweeping aside the defendant's argument (*PVA* v. *EBAE* 1996k: 401 TFH MO 20 Dec 1996: 16-17t).

On the third issue of obstructed sight line, the defendant had proposed two possible solutions.

The "no-stand" policy called for "education and enforcement" that would discourage customers in front of wheelchair seats from standing up. Signs would be posted describing the rule and employees would remind violators of its existence. The judge declared that, contrary to the aim of the ADA, this proposal involved an operational, not a design solution. The judge also maintained that the idea of seeking to keep only those spectators in front of wheelchair patrons from standing raised the "very real danger of subjecting wheelchair users to resentment and hostility" that "very likely [would] prevent them from fully enjoying the camaraderie and fellowship of a sporting contest or other event" (*PVA* v. *EBAE* 1996k: 403).

Under the "no sell" policy, seats in front of wheelchair patrons could be removed or withheld from sale. This tactic had much greater appeal to the court, which noted that a similar measure had been approved by the DOJ for the Olympic Aquatic Center in Atlanta, thereby suggesting that the policy complied with ADA regulations. Judge Hogan astutely pointed out, however, that disallowing the use of the seats in front of wheelchair spectators isolated those in wheelchairs, thereby constituting what he labeled a failure in vertical integration, though he believed that this was not an unreasonable loss and observed that front row spectators shared the same kind of separateness.

On the fourth issue, the court noted that most of the seats allocated to those in wheelchairs were not well dispersed; that they were, to use his term, "ghettoized" (*PVA* v. *EBAE* 1996k: 398). There was in particular an absence of an adequate number of places in the popular and expensive center court sections. This condition would have to be remedied before the MCI design could be found to be in compliance with ADA standards.

On the fifth issue, he ruled that the suite spaces, since they would not really be available for purchase by the general public, could not be counted in the number of required wheelchair-accessible locations, though each suite did need to provide the formula-required number of accessible locations.

When the MCI Center finally opened in December 1997, the PVA still found that compliance was not total, but it had not decided whether the remaining matters were important enough to allocate further resources to see that they were altered (Haggerty 1998).

The U.S. Court of Appeals, Washington, D.C. (July 1997)

Both parties took their legal grievances upward to the United States Court of Appeals, focusing almost exclusively on the meaning of the "lines of sight comparable" phrase in the DOJ guidelines. The opinion of the three-judge court, the first appeals court ruling in the country on the wheelchair access in sports arenas issue, was written by Circuit Judge Lawrence H. Silberman. It began by wrestling with the appellants' contention that DOJ regulations when properly read did not (and could not without public notice and the opportunity for comment) require equivalent sight lines for wheelchair attendees in regard to barriers created by standing patrons. The PVA, on the other hand, insisted that all (not just a certain percentage) wheelchair seating should have unobstructable sight lines (*PVA v. DCA* 1997c).

The Court of Appeals dug deeply into the background of the language employed in Standard 4.33.3, noting that it derived from a 1980 declaration of a private organization, the American National Standards Institute (ANSI) (Document A117.1-1980). The language subsequently appeared in rules issued in 1991 by the Architectural and Transportation Barriers Compliance Board (known as the Access Board), a group of 25 persons—13 appointed by the president and representatives of each of 12 government departments. The Access Board had proposed that wheelchair seating be "located to provide lines of sight comparable to those for all viewing areas" and pointed out that its wording "may not suffice in sports arenas or racetracks where the audience frequently stands." It therefore solicited comments on "whether full lines of sight over standing spectators ... should be required" (56 Fed. Reg. 2296, 2314 1991; and *PVA v. DCA* 1997c: 581 Opinion 01 Jul 1997: 4t). In July (1991), the Access Board issued a guideline that did not in any manner deal with the standing spectator issue, saying that it would touch upon that concern when it crafted guidelines for recreational facilities (56 Fed. Reg. 35,408, 35,440 [1991]), something that it never accomplished. The judge noted of all this "[a]lthough there is no indication that the words were intended to address sight lines over standing spectators, neither is there any evidence to the contrary" (*PVA* v. *DCA* 1997c: 583).

The DOJ had adopted the Access Board's wording verbatim in July 1991. In 1992, Irene Bowen, the deputy chief of the Public Access Section of the DOJ, in a speech to operators of major league baseball stadiums had said that "[t]here is no requirement [in the guidelines] of line of sight over standing spectators" (*PVA* v. *DCA* 1997c: 581). But in the December 1994 supplement to its *Americans with Disabilities Act Title III Technical Assistance Manual*, the DOJ published, without providing prior notice and without seeking public comment, an explicit interpretation of the "lines of sight comparable" requirement that stated: "[I]n assembly areas where spectators can be expected to stand during the event or show being viewed, the wheelchair locations must provide lines of sight over spectators who stand" (*PVA* v. *DCA* 1997c: 582).

The key question for the court was whether the DOJ legally could alter its position on the meaning of sight lines merely by revising its technical manual. The court granted that if the Access Board that originally had promulgated the guidelines had made the change this would have been an amendment requiring public notice and comment. But it argued (rather unconvincingly, we believe) that though the DOJ wholly adopted the Access Board regulation, this did not necessarily mean that it also agreed with the Access Board's statements about what it intended by the words or what later actions it meant to take in regard to the issue. The court declared that nothing prevented the DOJ from imposing a greater burden on those entities covered by its regulation (*PVA* v. *DCA* 1997c: 582). It then presumed that the DOJ had meant to do so, and that the sight-line requirement was not an amendment, but an explication of the DOJ's understanding of what was intended by its own rule. The contrary statement by the deputy chief of the Public Access Section of the Civil Rights Division of the Department of Justice was disposed of as "[a] speech of a mid-level official of an agency" and "not the sort of 'fair and considered judgment' that can be thought of as an authoritative departmental position" (*PVA* v. *DCA* 1997c: 587 Opinion Silberman 01 Jul 1997). Another court later also dismissed the statement, saying: "[p]ost-election remarks made by a mid-level official in a lame-duck administration at a convention of baseball stadium operators do not constitute a binding interpretation of agency regulations" (*ILR* v *OAC* 1997f: 737). Therefore, the later interpretation in the rules manual of the phrase "lines of sight comparable" was deemed not sufficiently distinct or additive to the regulation to require the procedural notice and comment[2].

Even though the judge thus dismissed a public comment by a high ranking relevant official in the agency charged with developing, interpreting and enforcing the regulations, he presumed that an employee of Ellerbe Becket could be responsible for the company:

> DeFlon was a vice-president (or perhaps senior vice-president) of Ellerbe Becket's sports design group. The knowledge that DeFlon obtained from his meetings with DOJ, in his capacity as Ellerbe Becket's representative, regarding legal requirements for designing arenas and stadiums ... is imputed as a matter of law to Ellerbe Becket as his employer (*ILR* v. *OAC* 1997f: 750).

Nonetheless, the court had the grace to obliquely acknowledge how far it had reached to arrive at its judgment:

> We admit the issue is not easy; appellants almost but do not quite establish that the Department significantly changed its interpretation of the regulation when it issued the 1994 technical manual. We conclude, finally, that the Department never authoritatively adopted a position contrary to its manual interpretation and as such it is a permissible construction of the regulation (*PVA* v. *DCA* 1997c: 587).

The Court of Appeals thus seconded Judge Hogan on the point of rule making.

The Court of Appeals also supported Hogan in his determination that "substantial compliance" with the regulations, rather than total compliance, was acceptable, so that a range of 78 percent to 88 percent of required wheelchair seating with unobstructable viewing was adequate (*PVA* v. *DCA* 1997c: 582, 588 Silberman Opinion 01 Jul 1997). This conclusion rested largely on what the court saw as some practical tensions between the line of sight requirement and the demand that wheelchair seating be dispersed throughout the arena.

The Court of Appeals also took a few swipes at the DOJ for failing to be helpful, despite what the court observed was its aggressive enforcement posture in similar cases (see, for example, *U.S.* v. *Physorthorac* 1996; Dunlap 1997). It found that Hogan had been "understandably exasperated" in regard to the government's failure to cooperate satisfactorily and that he had acted within his discretion in refusing to allow the government to submit a second amicus brief "purporting to refine arguments that could have been presented in its first [brief]" (*PVA* v. *DCA* 1997c: 589).

The appellate court decision remains binding only in its jurisdiction. Nonetheless, opinions, particularly of well-regarded courts, such as the D.C. Circuit, tend to be very carefully considered by other federal appellate tribunals at the same level.

Seeking to avoid that likelihood, and to avoid being hard hit financially in terms of revenue not realized, the defendants tried their luck with the U.S. Supreme Court, but in March 1998 that court let stand the appellate court ruling (*PVA* v. *DCA* 1998 USSC; Haggerty 1998).

CASE II: THE ROSE GARDEN, PORTLAND, OR

U.S. District Court, Portland, OR (November 1997)

The decision of the federal district court in Oregon in the case of Independent Living Resources (ILR) v. Oregon Arena Corporation (OAC) revisited in a new venue many of the issues raised in regard to the MCI Center, though with a strikingly different outcome on sight lines, the major matter of contention. In addition, the court addressed interesting further concerns including some gloss on conclusions reached by Judge Hogan in the MCI Center case.

The case had been filed on January 26, 1995. The plaintiff was a private, nonprofit group devoted to assisting persons with disabilities. Formed in 1957, the group had an operating budget of about $760,000 per annum derived from a variety of governmental and private sources. Its Board of Directors must have a majority of persons with disabilities (Ciesielski 1998).

We will not address each of the matters that came before the court but only several that had arisen in the MCI case or that are particularly important, interesting or concern matters that are key to our later discussion. Among the issues omitted in our review are the judge's ruling that arena suites had to provide visual alarms for the benefit of deaf spectators and that the accessibility requirement would not be met if wheelchair patrons would have to call in advance to let arena operators know that they were going to occupy a suite seat "as if they were bearers of a contagious disease" (*ILR* v. *OAC* 1997f: 764 Ashmanskas Opinion 12 Nov 1997: 115t).

The Rose Garden, the arena at the core of the controversy, was the home of the Portland Trail Blazers basketball team of the NBA and the Portland Winter Hawks of the Western Hockey League, and also the site for other kinds of entertainment.

Ellerbe Becket at first had designed this arena so that ambulatory companions would have seating space a row behind wheelchair patrons. Much more wary after the MCI judgment, the architectural firm had warned OAC to seek legal counsel before adopting its seating design because it had learned that four individuals in the Department of Justice had "interpreted the term 'next to' to mean side-by-side, not just in close proximity" (*ILR* v. *OAC* 1997f: 704 Opinion 12 Nov 1997: 11t). After some angry exchanges between the architects and the owners, side-by-side seating had been designed and, in a move that the judge presumed was born of desperation, OAC had tried to compensate for the loss of wheelchair spaces by placing 33 new wheelchair seats on Level 7, the proverbial "nosebleed" section, where there would be only 16 seats for ambulatory spectators, making, in the judge's eyes, a mockery of the ADA's dispersal requirement (*ILR* v. *OAC* 1997f: 705-706 Opinion 12 Nov 1997: 13-14).

The court devoted considerable attention to the question of whether companion seating needed to be "fixed" (the term used in the ADA regulations), that is, bolted into the floor, and the same make as the standard arena bowl seating or whether it could be padded Clairin folding chairs. The judge thought the folding chairs adequately fulfilled the requirement for seats for ambulatory companions of wheelchair customers; indeed, that the flexibility of folding chairs might prove useful to wheelchair patrons with special spatial needs. With some asperity, he noted: "The purpose of a companion chair is not to be 'fixed,' but to provide a seat for a companion" (*ILR* v. *OAC* 1997f: 725).

The judge deemed it unnecessary to litigate the comfort of the chair "given the number of hours spent seated on the bench ... if there is one thing that the average federal judge is qualified to do, it is to select a comfortable chair." Clairin chairs, it was officially declared, "were not significantly less comfortable than the standard seats used in the rest of the arena" (*ILR* v. *OAC* 1997f: 725).

Nor was the judge impressed by the argument that the folding chairs undercut the desire of wheelchair patrons to blend in with the audience and not feel conspicuous. "[T]he reality," he pointed out, "is that wheelchair patrons are not identical to ambulatory patrons; they use wheel-

Case Studies in Implementation of the ADA 221

chairs for locomotion ... the physical appearance of ... wheelchair locations necessarily will be different ..." (*ILR* v. *OAC* 1997f: 726). These rulings may reasonably be regarded as judicial common sense pretending to be an accurate reading of what the court regards as an unreasonable regulation.

The judge also revisited the line of sight issue that had gotten so much attention in the MCI litigation, noting that standing to cheer is encouraged in arenas. Radio advertisements for the MCI for example implored: "Fans, get on your feet and welcome your Washington Wizards" (Hagel n.d., p. 1). Wheelchair spectators at crucial moments of a game would see only the backs or backsides of those standing in front of them (*ILR* v. *OAC* 1997f: 732-733 Opinion). He dismissed the argument that children suffer the same difficulty, noting that their parents may hold them up above the crowd; besides, theirs is only a temporary impediment; they will grow taller during the coming years.

Revisiting the defendant's argument, one met earlier in the MCI litigation, that wheelchair users were demanding "preferential treatment," the judge resorted to biting sarcasm, observing that "the class of wheelchair users is always open to new members," and adding: "However, the court has not observed—and does not anticipate—a rush of volunteers choosing to have their legs amputated so that they may watch a basketball game from a wheelchair and thereby enjoy 'preferential' sight lines" (*ILR* v. *OAC* 1997f: 734 Opinion 12 Nov 1997: 56-57t).

Judge Ashmanskas swept away the defendant's argument that it was required only to provide equivalent seating to wheelchair spectators by noting that equivalent referred to function, not merely to physical conditions. If the defendant provided "identical" men's and women's rest rooms at the Rose Garden, the court observed, each containing two toilet stalls and seventeen urinals, could it then be argued that the facilities were comparable because of their physical similarity? He thought the same logic applied to ADA standards: what was done must assure functional equivalence.

The judge then entered into a laborious inventory of the actions (and, more notably, the inactions) of the Access Board and the DOJ on the matter of sight lines over standing spectators and concluded that the OAC was not required to provide lines of sight over standing spectators because there was no explicit or strongly implicit requirement for such in the existing regulations and the commentaries upon them (see in accord *Caruso* v. *Blockbuster*, 1997).

Finally, Judge Ashmanskas went into considerable detail to support his view that if the appellate court above him decided that there indeed was a legal requirement to provide adequate sight lines for wheelchair patrons seated behind ambulatory patrons—that is, if such a rule had been properly promulgated—then neither Ellerbe Becket nor OAC had a satisfactory basis, either in equity or otherwise, to be exempted from compliance with that requirement.

U.S. District Court, Portland, OR (March-April 1998)

Almost 100 outstanding issues, concerning particular aspects of the construction of the Oregon Arena were considered by Judge Ashmanskas in opinions delivered four and five months after his initial ruling. Each of the matters dealt with the question of whether the builders had met ADA requirements in regard to such things as protruding objects, knee clearance in tables, the obstruction of accessible routes and similar matters. Some were moot—they had been resolved to the satisfaction of the plaintiffs—while others were decreed to be up to the standard despite the reservations of the plaintiffs and still others were said to require additional actions. Typical was the plaintiffs' complaint that the center lines of two toilets were 19 inches from the side wall though the standards mandated 18 inches. The defendant insisted that the deviation was within "dimensional tolerance." The judge ruled that there had been no showing of what such tolerances might be and therefore the defendant had failed to meet its burden in regard to the center lines, and ordered changes.

The Mediation Settlement

The Portland case came to a conclusion when a settlement was hammered out between the contesting parties. The ILR negotiated successfully for places for 101 wheelchairs in the Oregon Arena. Judge Ashmanskas' elaborate dissection and sandbagging of the OAC's possible defenses if a higher court declared a necessity for adequate sight lines for all wheelchair patrons (*ILR* v. *OAC* 1997e: 785, Ashmanskas Opinion:157t) also apparently paid off: The defendants agreed to provide unobstructable sight lines for all wheelchair patrons. They also agreed not to put on public sale any wheelchair places unless they had not been reserved by the morning before an event. For their part, the plaintiffs abandoned their quest to have the suites retrofitted to provide adequate accommodations for wheelchairs (Ciesielski 1998).

DISCUSSION

What insights might we extract about public policy from our review of two sports arena court cases related to the ADA law that attempted to change the way society related to persons with disabilities?

We note that the American tripartite division of public policy into formulation, development, and implementation did not work smoothly. The ADA, as a legislative policy, has been criticized by at least one observer who notes that the statute was "an example of a badly drafted law" (Conrad 1998, p. 8; Condrey and Brudney 1998, p. 39). To make matters worse, as we have detailed, the agencies charged with developing guidelines at times did not do a particularly commendable job. Judges, who thereafter were saddled with complex adjudicative tasks, sometimes pointed out in annoyance that they had to fill in for the legislators and for the government agencies whose inaction or incomplete action left matters in limbo. And court rulings were not always clear, unequivocal, uncontroverted, or aligned with the objectives of the U.S. Congress.

The courts themselves did not have the resources to take on many of these tasks. They usually lack investigative resources and technical literacy in regard to ADA issues they must decide. Besides, they must resolve competing claims about a specific rule in terms of one instance (e.g. the MCI Center or the Portland Arena), though their decisions may be applied to a large universe of similar or idiosyncratic situations (Diver 1983).

But, persons with disabilities should not have to go to court to obtain compliance with a law. One side effect of the use of the courts and their adversarial system in the policy process was that a policy that was meant to have society integrate persons with disabilities cast them as adversaries.

These case studies call into question regnant theories which do not mesh satisfactorily with the data. There exists, for example, a very extensive and fiery literature that maintains that moneyed forces control the lawmaking and enforcement process and forge it in their own self-interest (see Kolko 1962; Zey 1998). As a British jurist put the matter: "Justice is a word that people in power use to give moral cover to what is really their own material interest" (Davenport-Hines 1995, p. 61). The wheelchair sight line cases indicate, however, that a powerful statute, with a strong moral thread, can provide a rock upon which strong corporate forces can be broken.

These lawsuits were filed by "social watchdog" organizations fighting for what they saw as the legally mandated rights of persons with disabilities. They were opposed by powerful corporate forces that had the resources to employ skilled attorneys to carry the case all the way to the U.S. Supreme Court. Their appellate briefs were marvels of sophisticated analysis, well researched, sedulously argued, and compelling. Even so, the courts ruled against them on several matters. The major issue with which the judges had to deal was simple enough: Did the law require assembly areas to be designed so that persons in wheelchairs would have unobstructed sight lines? But so straightforward a matter was subtly embedded in a number of moral, juridical, fiscal and other concerns, as we describe below.

First, the law had a strong and fairly clear moral force and agenda. It was the moral force undergirding this law, the idea of correcting a wrong, of providing opportunities to a set of citizens, we would argue, that played a highly significant part in the law having its effects. The ADA was primarily a powerful declaration of a legislative desire to move persons with disabilities into the mainstream. It provided direction for subsequent actions and decisions; without it policy would have become a compilation of small and minor actions. It played a highly significant part in the judge's rulings in favor of the plaintiffs, even when the opposing claims were rather balanced.

A good example was the manner in which the courts handled the public statement that unobstructed views were not required under the existing guidelines. The person who made the statement obviously occupied a position of considerable authority in the DOJ, backed by direct experience with the drafting and enforcement of the guidelines. Yet, judges Silberman and Ashmanskas dismissed the remark as an observation of an unimportant bureaucrat.

The moral issue was a good deal more subtle than the legal, but a close reading of the case materials shows how it came to impinge on matters of fact. In dealing with the question of whether the DOJ adopted the Access Board commentary, a matter of some legal importance, and after spending eight pages discussing the issue, Justice Ashmanskas brushes it all aside stating, "[t]he ADA was intended to usher in a new era for persons with disabilities" (*ILR* v. *OAC* 1997f: 743 Opinion 12 Nov 1997: 72t). Another striking instance was his stinging response (quoted earlier) to the defendants' point that unobstructed wheelchair views would provide more than equal treatment. Similarly, Judge Hogan proved particularly sensitive to the position of spectators in wheelchairs when he objected to a seating arrangement that would force those in front of wheelchair patrons to remain seated at all times. Such a situation, the judge declared, might result in resentment and hostility (clearly against the moral direction of the ADA) toward the wheelchair spectators and very likely would prevent them from enjoying the camaraderie and fellowship that often prevails at sports events.

A careful examination of the cases that arose from the wheelchair sight line issue might understandably lead an unbiased outsider to conclude that the outcome was more in the nature of the application by judges of principles of equity; that is, an attempt to achieve a desirable degree of social justice, and that the result, however commendable, was achieved at the expense of literal adherence to the letter of the law. A judge, so inclined, could readily have relied on all or any of these considerations to rule against the plaintiffs, as they at times did, or to send the lawmakers back to the drawing board.

The end result of the various court decisions on wheelchair-accessible seats guarantees that at least one set of persons with disabilities will be able to enjoy unimpeded the excitement of activities from which they often have been excluded by the unfriendly designs of arenas in which such activities are held. But it is the process by which that conclusion was reached that needs to be considered carefully, a task we set for ourselves in this paper.

Second, clarity in the formulation of public policy instruments such as laws, regulations, rules, and standards, can help in their implementation. Policy and lawmakers and regulators are in the best position to express what the end state should be, and what characteristics it would have. They need to consider how they will engage others in having their vision implemented. This, as we described with respect to the moral component, involves letting others know what the end state will be, not necessarily specifying in increasing detail what particular actions to take. Some policy makers recognize the problems with vagueness but seem unable to influence events:

The legislative history reveals that some members of Congress were uncomfortable with this [subsequent rule making and adjudication] feature of the ADA. Rep. Douglas protested that "Congress is abrogating it[s] constitutional duty by writing vague laws which must be clarified by the federal courts. *Our responsibility* is to write laws which can be clearly understood when reading them—not have another branch of government do our job" (H.R. Rep. No. 101-485{III} at 94 [1990], *reprinted at* 1990 USCCAN 511 (emphasis in original). His objections did not carry the day (*ILR v. OAC* 1997f: 737-738 Opinion 12 Nov 1997: 63-64t).

Vagueness in the framing of the law leads to government agency officials determining details. Agencies charged with the responsibilities of interpreting and enforcing may be lax or tardy in carrying out their responsibilities, as these cases show.

Vagueness also leads to people not knowing what action to take. Judges, the supposed legal experts, struggled with and even made contradictory rulings on matters of law. In legal terms, there was considerable doubt that the law required unobstructed sight lines as neither the original statute nor the initial guidelines specifically required it. A late incorporation of the sight line requirement into the standards had been done without regard to the Administrative Procedures Act's mandate of prior publication and comment. Lawsuits resulted, and the judges had to figure out the intent of the policy:

> When confronted with the first opportunity to interpret another statute of great breadth—the National Environmental Policy Act—Judge Skelly Wright observed that the judicial role in that instance was "to see that important legislative purposes, heralded in the halls of Congress, are not lost or misdirected in the vast hallways of the federal bureaucracy" (*Calvert Cliffs' Coordinating Committee, Inc.* v. *United States Atomic Energy Commission,* 449 F2d 1109, 1111 (DC Cir 1971) (*ILR* v. *OAC* 1997f: 745 Opinion 12 Nov 1997: 78t).

But judges chafe at the idea of having to surmise the intentions of the policy makers. Judge Hogan devoted some space to this issue in his memorandum opinion and reluctantly concluded: "Therefore, the Court is forced to step in and decide issues which would have been far better left to the politicians in the executive and legislative branches" (*PVA* v. *EBAE* 1996k: 394 TFH MO 20 Dec 1996: 2t). Judges having to make decisions to fill in gaps did so in a common sense way (as we have pointed out), not necessarily the way the lawmakers would have liked. Vagueness in the law led in the MCI case, according to one commentator, to mistakes in judgment in selection of "substantial compliance" instead of rebuttable assumption (Fritts 1998; see also Conrad 1997; Haggerty 1998). The courts, as the cases indicate, in grappling with the decisions tended to focus on matters of law.

The cases draw attention to the importance of placing mandates as specifically as tactically possible in the laws passed by legislative bodies. The uncertain nature of the requirements and the failure of the regulatory agency to establish adequate guidelines made it necessary for the parties to face the uncertain outcome associated with court litigation.

Third, the cases highlight the importance of clarifying priorities among competing interests. The judges struggled to balance demands for visibility, seating dispersal, and integration with the numerical requirement specifying how many wheelchair places were to be included. Judge Hogan, for example, points out that "[i]f anything, plaintiffs have argued that the seating is too integrated" (*PVA* v. *EBAE* 1996k: 394 TFH MO 20 Dec 1996: 20t). Ultimately, the judges had to bypass the law's mandate in order to reach what they perceived to be a workable compromise. The compromise, unfortunately did not help matters for those this law was supposed to help. The lesson is that at the drafting stage of statutes concerned persons must lobby to see that appropriate priorities are recorded as clearly as possible rather than defaulting to lawmakers, administrative agencies and the courts.

Fourth, courts are an important supplement in the policy process. The cases indicate the importance of recourse to the courts to obtain rights that otherwise might not be granted to persons with disabilities if the decisions were left to entrepreneurs or public administrative bodies. It provides a venue, limited though that might be in format and content, and problematic though it might be (as pointed out earlier) for the parties to press their case.

Fifth, in the United States, the policy makers after enacting the ADA neglected to consider incentives and encouragement for compliance to make the law operate as well as it might have. Primary among those entrusted with compliance with Title III were architects whose business it is to find creative solutions to "problems" such as access. But as public policy this law (especially Title III) made little effort to engage the professionals involved, the architects and builders, to encourage creativity by setting up design challenges, providing guidelines with scenarios of end conditions (see also Dunlap 1997); instead the law dictated solutions. Rather than bring out the best in the architects and builders, in some ways this law brought out their not-so-good side. They ended up fighting the law, even though they did not dispute its moral component.

Nor did the ADA policy makers consider the fiscal aspects of compliance and who must bear them. Rendering space and structures accessible for the full and equal enjoyment of persons with disabilities involves expenses that someone must bear. This cost was to be borne primarily by the owners of the public assembly building; little by way of financial assistance was offered except some tax savings. The defendants fought to avoid the considerable expenditure (and loss of revenue) associated with the inclusion of unobstructed sight lines in sports arena designs. But they were up against the extraordinarily powerful ethos that had characterized the ADA from its inception: It was to be interpreted as civil rights legislation and the cost of meeting its demands was not to be regarded as a significant nor overriding concern.

"The ADA's most troubling defect, the decision to saddle American businesses with the ADA's resulting costs," one writer has maintained, "represents a congressional error in discretionary policy formulation. ... Congress foisted the ADA's costs upon the business section without any plausible justification" (Bernard 1992, p. 41). The other side of this fiscal debate presumably would counter that the expenses to businesses were not that awful and that they represented no more than a reasonable and affordable outlay to make more certain that we all live in a decent society in which we can take pride.

There were several kinds of costs from the incorporation of requirements of the law. Although their accuracy is unconfirmed, these figures were provided:

> In its response to the court's questions, defendant asserted that it would have cost at least $20 million to incorporate sight lines over standing spectators into the Rose Garden, assuming such a requirement had been incorporated in the design parameters from the outset, and a much larger sum if the design had to be modified at a later date. These contentions border on the incredible ... (*ILR* v. *OAC* 1997f: 749 Opinion 12 Nov 1997: 85-86t footnote 65).

There was the perceived "cost" of lost revenue. For example, the provision of one wheelchair-accessible seat was seen as equivalent of five to ten ambulatory seats. One further cost was the expense associated with the lawsuit and lawyers' fees. Avoiding vagueness would have reduced later costly remediations.

Sixth, it is useful to factor in what we call "implementation inertia or loss" (see also Pressman and Wildavsky 1973). From the point of view of legislatively addressing the plight of persons with disabilities, and from the point of view of persons with disabilities who were the supposed beneficiaries, the effect of the policy was partially lost in the fighting of court cases and rulings, some of which were not in their favor. Their hopes were raised and dashed until several court cases later when the DOJ sued Ellerbe Becket and obtained a consent decree. The consent decree will affect future, but will not rectify already designed stadiums. Another example is "substantial compliance," which represents up to a 25 percent loss.

Seventh, there was not much use of "empathetic policy making." Beyond the broad declaration of the prevention of discrimination, there was little effort in the law to consider the frame of reference of persons with disabilities and how they would like to see accessibility. Nor were other frames of reference considered, such as the point of view or problems to be faced by the stakeholders: persons with disabilities and even architects, builders, and owners.

CONCLUSION

It is important to return, in concluding, to the quotation marshaled by Judge Hogan in his ruling in the MCI sports arena case: De Tocqueville (1868, p. 196), he noted, had observed more than 100 years ago that "there is hardly a political question in the United States that does not sooner or later turn into a judicial one." The cases we have reviewed testify to the veracity of De Tocqueville's insight. That the litigants spent considerable amounts of money to try to obtain a court decision in their favor testifies to their view that the issues were arguable and the outcome far from predetermined.

"Those who interpret the law in effect make the law," is the headline on a recent op-ed piece. The author, a lawyer-scholar, states categorically: "Law is political" and points out that "interpretation of the law is a function of who is doing the interpretation. Law is plastic, immensely manipulative." The Constitution and statutes become "a judge's Rorschach test" (DiMento 1999: C3). The cases related to the ADA bear this out, especially judge Ashmanskas' ruling trying to prompt a more liberal Court of Appeals in Oregon.

The conclusion of the op-ed piece corresponds with ours. Understanding how the law works, it declares, "urges us to enter in to the discourse on what our laws should be and how they should be implemented, rather than have them be made, interpreted and enforced by those who claim that they know the true meaning of law" (DiMento 1999: C3). Professor DiMento strongly supports the position, as does our material, that those who desire change are obligated not only to exert their influence at the lawmaking stage but equally, or more so, at the subsequent stages. The moral is that the election and selection of those who sit on the nation's judicial benches is at least as important as the choice of lawmakers.

More generally, the cases demonstrate that the stirring declaration of Senator Tom Harkin (1990), sponsor of the ADA legislation, that the measure was an "Emancipation Proclamation" for persons with disabilities is not necessarily a conclusive statement but rather the evocation of a hope. We know of Lincoln's call for freedom and equality for former slaves. The subsequent history, however, would have to see what came about as far short of a true emancipation. Here, the lesson seems similar: it will be a long and painful trail to ensure that the ethos of the ADA is satisfactorily translated into acceptable practice (Dole 1994). This paper has been prompted by a desire to examine in terms of one aspect of the ADA how this goal might best be achieved.

ACKNOWLEDGMENTS

We express our appreciation to Lawrence B. Hagel, deputy general counsel of the Paralyzed Veterans of America, for his kindness in supplying us with briefs, court decisions, and press releases regarding the sports arena litigation; to Ellerbe Becket for information; to the MCI Center; to Julia Gelfand and Daniela Pappada, indefatigable tracers of relevant research materials. We thank Dr. Barbara Altman, Dr. Sharon Barnartt and anonymous reviewers for their helpful comments on an earlier draft.

NOTES

1. Court case citation page numbers refer to published ones where available. Reference to typescript versions are indicated with a t following the page number.

2. In a recent case, another court of appeals declared that notice-and-comment were necessary (*Caruso* v. *Blockbuster* 1999).

REFERENCES

Altman, B.M. 1981. "Studies of Attitudes Toward the Handicapped: The Need for a New Direction." *Social Problems* 28(3): 321-333.
Bernard, T.H. 1992. "Disabling America: Costing Out the Americans with Disabilities Act." *Cornell Journal of Law and Public Policy* 2(1): 41-62.
Bowles, S. 1996. "Arena's Architect Sued by Justice Dept." *Washington Post*, (October 11): B7.
Ciesielski, T. 1998. Telephone interview (August 31, 1998).
Condrey, S., and J.L. Brudney. 1998. "The Americans with Disabilities Act of 1990: Assessing Its Implications in America's Largest Cities." *American Review of Public Administration* 28: 26-42.
Conrad, M. 1997. "Sports Arenas Grapple with Disabilities Act." *New York Law Journal* (May 30): 5, 7.
———. 1998. "Disabled-seat Pact May Unify Standards for New Stadiums." *New York Law Journal* (May 8): 8-10.
Davenport-Hines, R. 1995. *Auden*. London: Heinemann.
de Tocqueville, A. 1868. *Democracy in America*. Translated by Henry Reeves. New York, NY: Harper.
DiMento, J.F. 1999. "Those Who Interpret the Law in Effect Make the Law. *Orange County (CA) Register* (January 14): C3.
Diver, C. 1983. "The Optimal Precision of Administrative Rules." *Yale Law Journal* 93: 65-108.
Dole, B. 1994. "Are We Keeping Our Promises to People with Disabilities?" *Iowa Law Review* 79: 925-934.
Dunlap, D.W. 1997. "The Disabled Present New Hurdles for Architects." *New York Times*, (Sun., June 1)9: 1,4.
Eckstein, H. 1975. "Case Study and Theory in Political Science." In *Handbook of Political Science*, edited by F.I. Greenstein and N.W. Polsby. Reading, MA: Addison-Wesley.
Edelman, J.M. 1964. *The Symbolic Use of Politics*. Urbana, IL: University of Illinois Press.
Feagin, J.R., A.M. Orum, and G. Sjoberg (eds.). 1991. *A Case for the Case Study*. Chapel Hill, NC: University of North Carolina Press.
Fine, M., and A. Asch. 1988. *Women with Disabilities: Essays in Psychology, Culture, and Politics*. Philadelphia, PA: Temple University Press.
Forgey, B. 1997. "Cheers for an Arena." *Washington Post* (November 22): B1, B7.
Fox, D.M. 1994. "The Future of Disability Policy as a Field of Research." *Policy Studies Journal* 22: 161-167.
Fritts, J.C. 1998. " 'Down in Front!': Judicial Deference, Regulatory Interpretation, and the ADA's Line of Sight Standard." *Georgetown Law Journal* 86: 2653-2675.
Hagel, L.B. (n.d.) "Arenas are Failing the Disabled." *ESPN SportsZone*.
Haggerty, M. 1998. "Ruling on Handicapped Access at MCI is Affirmed." *Washington Post* (March 10): E1, E4.
Hahn, H. 1993. "The Potential Impact of Disabilities Studies on Political Science (as Well as Vice Versa). *Policy Studies Journal* 21: 740-751.

Harkin, T. 1990. "Our Newest Civil Rights Act: The Americans with Disabilities Act." *Trial* 26 (December): 56-61.
Hobbes, T. 1651. *Leviathan*. London: Andrew Crocke.
Kolko, G. 1962. *Wealth and Power in America*. New York: Praeger.
Manning, J.F. 1996. "Constitutional Structure and Judicial Deference to Agency Interpretations of Agency Rules." *Columbia Law Review* 96: 612-696.
Pressman, J.L., and A. Wildavsky. 1973. *Implementation*. Berkeley, CA: University of California Press.
Ragin, C., and H.S. Becker. (Eds.) 1992. *What is a Case?* New York: Cambridge University Press.
Winston, S. 1997. "Accessibility: Disability Law Tests Architects. *ENR (Engineering News-Record)* (October 13): 239: 10.
Zey, M. 1998. "Embeddedness of Interorganizational Corporate Crime in the 1980s: Securities Fraud of Banks and Investment Banks." Pp. 111-159 in *Research in the Sociology of Organizations: Deviance in and of Organizations*, edited by P.A. Bamberger and W.J. Sonnenstuhl, Vol. 15. Stamford, CT: JAI Press.

CASES

Baltimore Neighborhoods, Inc., v. Rommel Builders, Inc. (1998). 3 F. Supp. 2d 661 (D. MD).

Caruso v. Blockbuster-Sony Music Entertainment Centre (1997). 968 F. Supp. 210 (D.N.J.)

Caruso v. Blockbuster-Sony Music Entertainment Centre (1999). *U.S. Lawweek*, p. 9-10. http://lw.bna.com/plweb-cgi/fastweb?getdoc+view6+lawweek+1725+0++

Independent Living Resources v. *Oregon Arena Corporation* (ILR v. OAC) (1995). Deposition of Gordon Wood, (USDC DOR CV95-84-AS Dep 04 Oct 1995) Sum: 982 F. Supp. 753-754.

Independent Living Resources v. *Oregon Arena Corporation* (1997a). Plaintiff's Complaint (USDC DOR CV95-84-AS 12 Nov 1997). 982 F.Supp. 698 (D. OR.) 706-770.

Independent Living Resources v. *Oregon Arena Corporation* (1997b). Defendant's Motion to Dismiss. (USDC DOR CV95-84-AS 12 Nov 1997) 982 F.Supp. 698. (D. OR.) 770-784.

Independent Living Resources v. *Oregon Arena Corporation* (1997c). Amicus Curiae brief by brief by U.S.A. DOJ (USDC DOR CV95-84-AS 12 Nov 1997). 982 F.Supp. 698 (D. OR.)

Independent Living Resources v. *Oregon Arena Corporation* (1997d). Plaintiffs' Rebuttal to Defendants' Motion to Dismiss. (USDC DOR CV95-84-AS 12 Nov 1997). 982 F.Supp. 698 (D. OR.).

Independent Living Resources v. *Oregon Arena Corporation* (1997e). Judge Ashmanskas' Order (USDC DOR CV95-84-AS Order 12 Nov 1997). 982 F.Supp. 698-786 (D. OR.).

Independent Living Resources v. *Oregon Arena Corporation* (1997f). Judge Ashmanskas' Opinion (USDC DOR CV95-84-AS 12 Nov 1997). 982 F.Supp. 698-786 (D. OR.).

Independent Living Resources v. *Oregon Arena Corporation* (1998a). Findings of Fact and Conclusions of Law. (USDC DOR CV95-84-AS FFCL 26 Mar 1998). 1 F. Supp. 2d 1124-1159.

Independent Living Resources v. *Oregon Arena Corporation* (1998b). Supplemental Findings of Fact and Conclusions of Law. (USDC DOR CV95-84-AS SFFCL 08 Apr 1998). 1 F. Supp. 2d 1159-1173.

Inman v. *Binghamton Housing Authority* (1957). 3 N.Y.2d 137, 143 N.E.2d 895, 164 N.Y.S.2d 699.

Johanson v. *Huizenga Holdings Inc.* (1996). Memorandum of Law of Amicus Curiae United States in Opposition to Defendants Ellerbe Becket Architects and Engineers, Inc.'s Motion to Dismiss, (USDC SDFL 96-7026-CIV-Gonzalez 31 Oct 1996).

Johanson v. *Huizenga Holdings Inc.* (1997). Order/Opinion. (USDC SDFL 96-7026-CIV-Gonzalez 27 Jan 1997). 963 F. Supp. 1175 (S.D. Fl.).

McPherson v. *Buick Motor Co.*, (1916). 217 N.Y. 382, III N.E. 1050.

Paralyzed Veterans of America (PVA) v. *Ellerbe Becket Architects & Engineers P.C.* (EBAE) (1996a). Plaintiff's Complaint for Declaratory Judgment and Injunctive Relief. (USDC DC 96CV01354 14 June 1996) 945 F.Supp. (D.D.C. 1996) sum: 1-3.

Paralyzed Veterans of America v. *Ellerbe Becket Architects & Engineers, P.C.* (1996b) Defendants' Motion to Dismiss Counts I, II & III, (USDC DC 96CV01354 (TFH), DMD 28 June 1996) 945 F. Supp. (D.D.C. 1996) sum: 1-3.

Paralyzed Veterans of America v. *Ellerbe Becket Architects and Engineers, P.C.* (1996c). Judge's Preliminary Temporary Injunction. (USDC DC 96CV0135) 945 F. Supp 1 (D.D.C. 1996) or 950 F. Supp. 389

Paralyzed Veterans of America v. *Ellerbe Becket Architects & Engineers, P.C.* (1996d). 950 F. Supp. 393 (USDC DC), Affirmed 117 F3d (1997) 579-589.

Paralyzed Veterans of America v. *Ellerbe Becket Architects & Engineers, P.C.* (1996e). The American Institute of Architects' Brief Amicus Curiae in support of EBAE & EBI (USDC DC 96CV1354 (TFH), AIA AC 10 July 1996) 945 F. Supp. 1 (D.D.C. 1996).

Paralyzed Veterans of America v. *Ellerbe Becket Architects & Engineers, P.C.* (1996f). Amicus Brief by DOJ (USDC DC 96CV01354 DOJ AC) 945 F. Supp. (D.D.C. 1996) sum:2.

Paralyzed Veterans of America v. *Ellerbe Becket Architects & Engineers, P.C.* (1996g). Judge Thomas F. Hogan's Order USDC DC 96CV01354 O 19 July 1996) 945 F. Supp. (D.DC 1996) 2-3.

Paralyzed Veterans of America v. *Ellerbe Becket Architects & Engineers, P.C. (*1996h). Judge Thomas F. Hogan's Order/Opinion (USDC DC 96CV01354 O 21 Oct. 1996) 950 F. Supp. (D.DC 1996) sum: 389.

Paralyzed Veterans of America v. *Ellerbe Becket Architects & Engineers, P.C.* (1996i). Judge Thomas F. Hogan's Memorandum Opinion explanation of Order 21 Oct. 1996 (USDC DC 96CV01354 TFH MO 20 Dec 1996) 950 F Supp (DDC 1996) 389-393.

Paralyzed Veterans of America v. *Ellerbe Becket Architects & Engineers, P.C.* (1996j). Judge Thomas F. Hogan's Order (USDC DC TFH O 20 Dec 1996) 950 F Supp (D.DC 1996) 405-406.

Paralyzed Veterans of America v. *Ellerbe Becket Architects & Engineers, P.C.* (1996k). Judge Thomas F. Hogan's Memorandum Opinion explanation of Order 20 Dec.

1996 (USDC DC 96CV01354 TFH 20 Dec 1996). 950 F Supp (D.DC 1996) 393-405.
Paralyzed Veterans of America v. *Ellerbe Becket Architects & Engineers, P.C.* (1996l). Judge Thomas F. Hogan's Order (USDC DC 96CV01354 TFH O 19 Feb. 1997 (typescript).
Paralyzed Veterans of America v. *D.C. Arena L.P.* (1997a). Appellants filing (USCOA D.DC. Cir. 97-7005 01 July 1997). 117 3d. (DC cir. 1997) 579.
Paralyzed Veterans of America v. *D.C. Arena L.P.* (1997b). Appellees filing (USCOA D.DC. Cir. 97-7005 01 July 1997) 117 3d. (DC Cir 1997) 579.
Paralyzed Veterans of America v. *D.C. Arena L.P.* (1997c). Opinion - Judge Lawrence H. Silberman (USCOA D.DC. Cir. 97-7005, O 01 July 1997) 117 F. 3d. (DC Cir 1997) 579-589.
Paralyzed Veterans of America v. *D.C. Arena, L.P.* (1998). 117 F3d 579 (USSC *Certiorari Denied*, 118 S.Ct. (1998) 1184 523 U.S. 1003.
United States v. *Days Inn of America, Inc.*, (1998). 997 F. Supp. 1080 (C.D. ILL).
United States v. *Ellerbe Becket, Inc.*, (1996) Complaint (USDC DMN 4-96-995 10 Oct 1996).
United States v. *Ellerbe Becket, Inc.*, (1997a) 976 F. Supp. (D. MINN) 1262.
United States v. *Ellerbe Becket, Inc.*, (1997b). Amici Curiae the American Institute of Architects & the Associated General Contractors of America's in support of Ellerbe Becket, Inc.'s Motion to Dismiss, (USDC DMN 4-96-995 20 Jan. 1997) 976 F.Supp. 1262 (USDC DMN).
United States v. *Ellerbe Becket, Inc.* (1998). Consent Order. (USDC DMN 4th 4-96-995 27 April 1998).
United States v. *Physorthorad Associates*, J. Wylie Bradley, and Bradley Chambers and Frey (1996). Consent Order (USDC DPA 4: CV-96-1077 CO 25 Jun 1996).